Spatial Information Systems

Chief Editor

P. A. Burrough

General Editors

M. F. Goodchild · R. A. McDonnell
P. Switzer · M. Worboys

Spatial Information Systems

Environmental Modeling with GIS
Edited by Michael F. Goodchild, Bradley O. Parks, Louis T. Steyaert (1994)

Managing Geographic Information System Projects
William E. Huxhold and Allan G. Levinsohn (1995)

Anthropology, Space and Geographical Information Systems
Edited by Mark Aldenderfer and Herbert D. G. Maschner (1996)

GIS County User Guide: Laboratory Exercises in
Urban Geographic Information Systems
*William E. Huxold, Patrick S. Tierney, David R. Turnpaugh,
Bryan J. Maves, and Kevin T. Cassidy (1997)*

Principles of Geographical Information Systems (2nd Edition)
*Peter Burrough and Rachael McDonnell (1998)
Paperback*

Land Administration
Peter Dale and John McLaughlin (2000)

Information Synthesis for Mineral Exploration
Goucheng Pan and Deverle P. Harris (2000)

GeoInformation Infrastructure
Richard Groot and John McLaughlin (2000)

Qualitative Spatial Change
Anthony Galton (2000)

Travel by Design: The Influence of Urban Form on Travel
Marlon Boarnet and Randall C. Crane (2001)

Spatial Evolutionary Modeling
Roman Krzanowski and Jonathan Raper (2001)

Geographic Information Systems for Transportation:
Principles and Applications
Harvey J. Miller and Shih-Lung Shaw (2001)

Environmental Systems

A Multidimensional Approach

ANTONIO S. CAMARA

OXFORD

UNIVERSITY PRESS

OXFORD
UNIVERSITY PRESS

Great Clarendon Street, Oxford OX2 6DP

Oxford University Press is a department of the University of Oxford.
It furthers the University's objective of excellence in research, scholarship,
and education by publishing worldwide in

Oxford New York

Auckland Bangkok Buenos Aires Cape Town Chennai
Dar es Salaam Delhi Hong Kong Istanbul Karachi Kolkata
Kuala Lumpur Madrid Melbourne Mexico City Mumbai Nairobi
São Paulo Shanghai Singapore Taipei Tokyo Toronto

and an associated company in Berlin

Oxford is a registered trade mark of Oxford University Press
in the UK and in certain other countries

Published in the United States
by Oxford University Press Inc., New York

British Library Cataloguing in Publication Data
Data available

Library of Congress Cataloging in Publication Data
Data available

ISBN 0-19-874267-3

1 3 5 7 9 10 8 6 4 2

Typeset in Ehrhardt MT
by SNP Best-set Typesetter Ltd., Hong Kong
Printed in Great Britain
on acid-free paper by
Biddles Ltd, Guildford and King's Lynn

To the memory of my parents, Manuel and Beatriz

To the future of Guida, Manuel, and Luis

Preface

This book is about initiatives to bring environmental management, engineering, and science to the multimedia age. They integrate old with new methods for collecting, storing, modelling, representing, and making available environmental information on the Internet.

In the World Wide Web (WWW or the Web), the fastest growing segment of the Internet, environmental information is increasingly multidimensional: it includes text, audio, graphics, photos, and video data types. As a result, environmental monitoring experts have to become familiar with multimedia acquisition and processing tools. Environmental information systems are now part of Internet based infrastructures. These provide access to maps, satellite images, aerial, and ground photos, geo-referenced video, audio, three-dimensional models, and alphanumerical data.

The wealth of multimedia information is facilitating the creation of a new generation of environmental models. They include traditional and new simulation methods coupled with visualisation and interaction techniques originally developed for other areas, such as the entertainment industry. Open access to environmental studies is now mandatory in many countries. The communication with the public is also gaining from the availability of information in the Web as reported in this book.

This book is directed to environmental professionals and students who want to have access to a compact presentation of multidimensional information-handling techniques. The approach, in writing it, was to combine the introduction of fundamental concepts with a guide to related expanded views. Web links and the Reference sections provide these views.

The book's Web site at http://gasa.dcea.fct.unl.pt/camara/index.html (username ASCAMARA, password ENVSYS) offers direct access to the suggested Web resources. It also includes demonstrations of illustrative applications of multimedia and virtual reality techniques to environmental problems. Web based student projects are proposed in the book's Web site. They are divided into three groups:

- Projects stressing fundamental concepts, which include the development and use of software applications.

- Topics for exploration.
- Proposals for collaborative projects that may be created by students (and other interested groups) around the world.

I have attempted to select Web links that seem to have a low probability of becoming broken in the near future. However, Web sites are, almost by nature, ephemeral. The proposed links will be updated periodically at the book's Web site (see above). This site also includes additional information and updates on student projects and reports on new developments in the areas covered in the book.

This book is the result of my research and consulting activities during the past ten years. In today's Internet age, information about these activities can be obtained by visiting the corresponding Web sites. My research group activities are annually displayed at summer seminars. These can be accessed at http://gasa.dcea.fct.unl.pt.

For the last seven years, I was a consultant to Parque Expo98. This firm organised Expo98, the World Exhibition held in Lisbon. I tempered my research on virtual reality with the engineering solution of very real pollution problems (http://www.parquedasnacoes.pt/in/ambiente/default.asp). I was also involved in the development of tools to communicate environmental problems to the public and media.

Finally, I was very proud to be involved in the development of SNIG, Portugal's infrastructure for spatial information (http://snig.cnig.pt). In 1548, the Royal Court organised a discussion on the strategic development of Portugal. The major question was whether Portugal should pursue knowledge or wealth. The courtiers voted for wealth and Portugal has suffered ever since. SNIG embodies, about 450 years later, a needed inversion.

In virtual reality (and in computer graphics in general), closer objects are always represented in more detail than those located farther. I almost naturally followed this principle when writing the book. My own experiences are its core. They come mainly from the above-mentioned projects. The 1998–9 academic year spent at the Massachusetts Institute of Technology provided additional perspectives. I apologise for the inevitable omissions and welcome the reader's feedback at asc@mail.fct.unl.pt.

A.S.C.

Lisbon, Portugal
February 2001

Acknowledgements

This book is the result of an invitation from Peter Burrough, who offered me the opportunity to write on the application of interactive techniques to environmental systems.

Most of the material presented here reflects the work developed by my students especially at the Environmental Systems Analysis Group (GASA) at the New University of Lisbon (UNL), during the last ten years. I especially thank Alexandra Fonseca, Ana Pinheiro, Antao Almada, Cristina Gouveia, Edmundo Nobre, Eduardo Dias, Fernando Lobo, Francisco Ferreira, Inês Sousa, Joana Hipólito, João Pedro Fernandes, João Pedro Silva, Joaquim Muchaxo, Jorge Nelson Neves, José Miguel Remédio, Julia Seixas, Maria João Silva, Pedro Gonçalves, and Teresa Romão for their contributions. At GASA, I am also particularly grateful for the support of Conceição Capelo, our administrative assistant, and Jorge Imaginario, our computer technician.

That work has been supported by the National Centre for Geographic Information (CNIG), Luso-American Foundation (FLAD), Ministry of Science and Technology (MCT), and Ministry of the Environment and Natural Resources (MARN). I am also most grateful for the personal support provided by Rui Gonçalves Henriques (President of CNIG) and Charles Buchanan (Vice-President of FLAD).

Most of the complementary research for this book was carried out during sabbatical leave at the Massachusetts Institute of Technology (MIT) with the financial support of FLAD, MCT, and UNL. David Marks was my host and I am thankful for his support and patience. Once again (I was a post-doctoral graduate with him in 1983) he opened up the MIT intellectual 'candy store' for me. I also thank Pete Loucks at Cornell University for his helpful criticism of earlier versions of the book.

I am grateful for the help and support provided by my wife Guida and son Manuel, both Web experts. Luis, my younger son—with his dangerous attraction to the wrong computer keys—forced me to back-up my work constantly and safely arrive at publication.

Finally, I thank the editorial staff of Oxford University Press who greatly improved the original manuscript.

Contents

List of Abbreviations

ACM	Association of Computing Machinery
ACS	American Chemical Society
ADSL	Asymmetric Digital Subscriber Line
AOL	America On Line
AR	Augmented Reality
ASA	American Statistical Association
ASP	Active Serve Page
ASPRS	American Society of Photogrammetry and Remote Sensing
AVHRR	Advanced Very High Resolution Radiometer
AVIRIS	Airborne Visible Infrared Imaging Spectrometer
BITS	Browsing in Time and Space
BOD	Biochemical Oxygen Demand
CASI	Compact Airborne Spectrographic Imager
CAVE	CAVE Automatic Virtual Environment
CEN	Comité Européen de Normalisation
CEO	Centre for Earth Observation
CERN	Conseil Européen de Recherche Nucléaire
CGI	Common Gateway Interface
CHAID	CHI Square Automatic Interaction Detector
CIESIN	Centre for International Earth Science Information Network
CMY	Cyan, Magenta, and Yellow
CNIG	National Centre for Geographical Information (Portugal)
COM	Component Object Modelling
DCT	Discrete Cosine Transform
DEM	Digital Elevation Model
DFT	Discrete Fourier Transform
DHTML	Dynamic Hypertext Markup Language
DIS	Distributed Interactive Simulations
DIVE	Distributed Interactive Virtual Environment
DML	Data Manipulation Language
DSS	Decision Support Systems
DTD	Document Type Definition
DTM	Digital Terrain Model
DWT	Discrete Wavelet Transform
EDA	Exploratory Data Analysis
EEA	European Environment Agency

EIA	Environmental Impact Assessment
EIONET	European Information and Observation Network
EO	Earth Observation
EPA	Environmental Protection Agency (USA)
ERIN	Environmental Research Information Network (Australia)
ESMI	European Spatial Metadata Infrastructure
ESRI	Environmental Systems Research Institute
EWSE	European Wide Service Exchange
FFT	Fast Fourier Transform
FGDC	Federal Geographic Data Committee
FLAD	Luso-American Foundation
FT	Fourier Transform
FTP	File Transfer Protocol
GASA	Environmental Systems Analysis Group, UNL
GASP	Globally Accessible Statistical Procedures
GDDD	Geographical Data Description Directory
GEO	Geosynchronous Earth Orbit
GIF	Graphic Interchange Format
GIS	Geographical Information Systems
GMD	National Research Center for Information Technology (Germany)
GOES	Geostationary Operational Environmental Satellite
GPS	Global Positioning System
HMD	Head-Mounted Display
HRTF	Head-Related Transfer Function
HSV	Hue, Saturation and Value
HTML	Hypertext Mark-up Language
HTTP	Hypertext Transfer Protocol
IAIA	International Association for Impact Assessment
ICMP	Internet Control Message Protocol
ICQ	I Seek You
IDEAS	Integrated Decision Aiding Simulator
IDL	Interactive Data Language
IEEE	Institute of Electrical and Electronics Engineering
IGBP	International Geosphere-Biosphere Program
IIS	Internet Information Server
IP	Internet Protocol
IRC	Internet Relay Chat
ISO	International Organization for Standardization
IWT	Inverse Wavelet Transform
JPEG	Joint Photographic Expert Group
LCD	Liquid Crystal Display
LEOs	Low Earth Orbits
LIDAR	Light Detection and Ranging
LOD	Level-Of-Detail
LZW	Lempel–Ziv–Welch
MARN	Ministry of the Environment and Natural Resources (Portugal)

MCT	Ministry of Science and Technology (Portugal)
MEL	Master Environmental Library
MIDI	Musical Instrument Digital Interface
MIME	Multi-Purpose Internet Mail Extensions (MIME)
MIT	Massachusetts Institute of Technology
MOOs	Multi-User Object Oriented
MPEG	Motion Pictures Expert Group
MUCK	Multi-User Collective Kingdom
MUDs	Multi-User Domains
MUSEs	Multi-User Shared Environments
MUSHEs	Multi-User Shared Hallucinations
NAPP	National Aerial Photography Program (USA)
NASA	National Aeronautics and Space Administration (USA)
NCGIA	National Centre for Geographic Information and Analysis
NCSA	National Centre for Supercomputing Applications (USA)
NIMA	National Imagery and Mapping Agency (USA)
NOAA	National Oceanographic and Atmospheric Administration (USA)
NSDI	National Spatial Data Infrastructure
NURBS	Non-Uniform Rational B-Spline
OBDC	Open Database Connectivity
ODL	Object Definition Language
ODMG	Object Database Management Group
OGC	Open GIS Consortium
OLE	Object Linking and Embedding
OQL	Object Query Language
OSI	Open Systems Interconnection
PbD	Programming by Demonstration
PbR	Programming by Reproduction
PDU	Protocol Data Unit
PRE	Product Ecology Consultants
PRN	Pseudo-Random Noise
RADAR	Radio Detection And Ranging
RAVI	Dutch Advisory Council on Spatial Information
RDBMS	Relational Database Management Systems
RGB	Red, Green, and Blue
SA	Selected Availability (time and orbit parameter errors in GPS)
SETI	Search for Extra-Terrestrial Intelligence
SGI	Silicon Graphics
SIGCHI	Special Interest Group on Computer Human Interaction (ACM)
SIGGRAPH	Special Interest Group on Computer Graphics (ACM)
SMTP	Simple Mail Transport Protocol
SNIG	National System for Geographical Information (Portugal)
SPARC	Space Physics and Aeronomy Research Collaboratory
SPIE	International Society for Optical Engineering
SQL	Structured Query Language
STM	Scanning Tunnelling Microscope

SVG	Scalable Vector Graphics
TCP	Transmission Control Protocol
TIN	Triangular Irregular Network
UARC	Upper Atmospheric Research Collaboratory
UDP	User Datagram Protocol
UNEP	United Nations Environment Program
UNL	New University of Lisbon
URL	Universe Resource Locator
USGS	United States Geological Survey
VE	Virtual Environment
VR	Virtual Reality
VRML	Virtual Reality Modelling Language
W3C	World Wide Web Consortium
WIM	World in Miniature
WVD	Weighed Voronoi Diagrams
WWW	World Wide Web
XML	Extensible Markup Language

List of Figures

List of Tables

1

Introduction

1.1 The Purpose of this Book

Environmental problems result from a difference between the existing and the desired values for variables that describe an environmental condition. For policy-makers, environmental problems exist, most of the time, only when dramatic events prompt media news and, as a result, a public interest. With the gradual decline of the media attention, the solution to the problem becomes less pressing. Sometimes, as pointed out by Downs (1972), there may be a long period with occasional resurgence of public interest and this may result in a solution.

Traditionally, environmental problems were perceived as being local and regional. Newson (1992) argues that the photograph 'Earthrise', captured from the Moon by the Apollo II crew in July of 1969, jump-started the awareness for the global problems the planet is facing today.

Environmental managers, engineers, and scientists routinely produce technical reports to address solutions for environmental problems. These solutions are presented as studies, projects, or plans. They also often present non-technical summaries that are subject to public consultation. Today, a growing number of these documents are published on the Internet as World Wide Web (WWW or Web) documents.

The Web is already a cost-effective platform for collaboration, decision support, and dissemination of information and participation in countries with a high level of Internet access and low communication costs such as the United States. Today, Europe and other parts of the world are developing a similar penetration of this medium. In general, the Web will become increasingly more attractive with the new Internet generations, and the availability of higher bandwidth and processing power. It is anticipated that the number of people connected to the Internet, in the developed world, will be a significant percentage of the total population in the near future as the convergence of television, computing, and telephony progresses.

This book is about existing initiatives to bring environmental management, engineering, and science to the Internet age or more generally to the multimedia age. It will also report on developments that may change the traditional methods of collecting, processing, and presenting environmental information. These developments are related to the availability of multidimensional data types (text, graphics, still images, video, and sound) to assist environmental modelling and visualisation.

The starting point is the hypothesis that most of the significant data, tools, and documents produced, for local, regional, and global environmental problems are currently available as Web sites. Many of these sites will include multidimensional data types. One can anticipate that these sites will be of two types:

- Sites mainly directed towards professionals and researchers.
- Sites mainly directed towards politicians, general public, and media.

The professional and research sites will include:

- Database-backed Web sites for individual environmental projects or initiatives.
- Information infrastructures, where producers make available their data and users can access (query) several distributed databases simultaneously.
- Modelling tools, such as simulation and other decision support tools, using data from the information sites.
- Interactive visualisation tools to facilitate and/or enhance data mining, database accessing (querying), and the operation of models.

Both the infrastructures and tools are already changing with the nature of the data and the increasing bandwidth and processing power. The availability of remote high-resolution sensors and ground sensors providing images, photographs, and video are adding new data types to information systems that used to be strictly alphanumerical.

The modelling and visualisation tools may be obtained by developing them in-house, as downloadable components to one's own site or as a service provided by a third-party site. The use of still images and video is enabling the integration of realistic representations in environmental studies, which were traditionally based on symbolic models. As shown in this book, a realistic interactive movie-based simulation can already replace a simple numerical environmental model.

The potential user population is also changing profoundly. New generations come from a videogame culture complemented with extensive chatroom participation. Thus, it can be expected that non-professional sites will be based on:

- *Interactive movie sites*. These sites will follow the videogame model and will be used to present environmental studies to decision-makers, media, and the general population using virtual environments. They will also be used for environmental education. These sites may be used in isolation or in association with community-based sites to foster citizen collaboration.

There are already Web sites and research projects either falling under these categories or demonstrating their feasibility. The production and use of multimedia Web sites are thus becoming integral elements of the professional toolkit of the environmental manager, engineer, or scientist. In fact, it can be argued that environmental professionals will add new roles such as:

- *Sensor managers*. With the expected increase in the availability of high-resolution sensors, environmental data monitoring may change dramatically. Environmental professionals will have to become proficient in sensor technology and their use. They will also need to understand multidimensional data types.
- *Information architects*. Many environmental professionals are already trained in databases and geographical information systems (GIS). They need to know how to develop Web sites backed by such systems. They will also have to understand about systems that handle distributed databases and GIS such as the information infrastructures already available around the world.
- *Tool-builders*. With new data types and increasing processing power, environmental models, and interactive visualisation tools are changing. New tools are required and so are their builders.
- *Interactive movie directors*. Environmental professionals structure their paper documents and some already do it for their Web sites. Structuring the multimedia sites that will present environmental studies to the media, public, and decision-makers will be similar to directing interactive movies.
- *Community moderators*. The sense of community attracts most young users to the Internet. Channels in chatrooms divided by topics are examples of the artificial communities that exist today. The creation and moderation of such communities will be essential to promote public collaboration. It may become a common activity in those professions related to the environment.

This book attempts to unite the methods and technologies which may help to foster these new roles.

1.2 Environmental Problems

Environmental problems range from natural disasters to the ecological design of new products (Box 1.1). Web sites relying on multidimensional data can be developed to address any environmental problem (such as traditional reports do).

The environmental problems considered in this book are however limited to the ones that can be represented spatially (on a map, aerial photograph, or satellite image). These include water, soil, air, noise, fauna, vegetation, geomorphology, and landscape and climate problems. Goudie (1986) provides a comprehensive description of the nature of such problems as a result of human intervention.

The book is not about these environmental problems *per se* but rather on available methods and technologies to represent them and contribute to their solution using Web sites. These sites are viewed here as extensions of traditional decision support systems (DSS) oriented for environmental problems such as the ones discussed in Loucks and Costa (1991), Watkins and McKinney (1995), Lein (1997), and Harmancioglu *et al.* (1998). As DSS, they are backed by databases and often rely on models to support decisions.

Box 1.1 *Industrial Ecology*

The goal of industrial ecology, of increasing interest to environmental professionals, is to have industry emulate natural ecosystem characteristics (Fiksel, 1996) such as:

- Efficiency and productivity are held in dynamic balance.
- There is no such thing as waste. Nutrients from one species are derived from the death of another.
- The natural system is dynamic and information-driven and the identity of ecosystem players is defined in process terms.
- Cooperation and competition are interlinked and in equilibrium.

Industrial ecology practices include pollution prevention strategies by process change and ecological design (Richards, 1997). Web sites related to pollution prevention may be found in the US Environmental Protection Agency (EPA) site mentioned in Section 1.6. This site is a repository for initiatives on pollution prevention, compliance assurance, and enforcement information. A list of eco-designed products may be seen at the Massachusetts Institute of Technology (MIT) Gallery site. This Product Ecology Consultants (PRE) site includes information on life-cycle assessment, designer tools, and eco-indicators.

Industrial ecology may potentially apply many of the methods covered in this book.

1.3 Introducing Multidimensional Web Sites

The World Wide Web is the platform for development and exploration of multi-dimensional environmental sites. Two other key technologies are multimedia and virtual reality. Multimedia data and tools enable the handling of images, video, and sound in Web-based environmental decision support systems. Virtual reality technologies provide interactive visualisation tools for three-dimensional representations of environmental problems.

1.3.1 World Wide Web

The World Wide Web, developed at Conseil Européen pour La Recherche Nucléaire (CERN) in 1991 (Berners-Lee, 1996), is the fastest growing sector of the Internet. Its success is due to the easy authoring of documents with appropriate editing tools and the inexpensive access it provides to communication, research, and advertising.

The Web is based on the hypertext concept which is text that does not need to be read linearly. Hypertext is a loose web of programmed interconnections, where the author provides links in the information that points to other relevant portions of the text (Nielsen, 1994). Hypermedia extends the concept of hypertext to include static and dynamic images, and sound.

HyperText Mark-up Language (HTML) is the data format for interchange of hypertext between WWW client and server. Web pages are written in HTML and can be read on many different computer platforms. Any HTML document may include multidimensional data types, such as text, graphics, animation, video, and sound, and provide access to databases and geographical information through a variety of alternative methods. It can also enable the exploration of virtual environments. As a result, the Web is an integrating system for components of a multidimensional approach to environmental decision support systems.

Section 1.5 includes a directory of sites that are illustrative of the potential applications of the WWW to environmental problems: digital libraries, institutional sites, and information sources for environmental studies. Addresses of sites for specific environmental problems are included in Chapter 2.

A source guide to Web site development and maintenance is included in Appendix 1. Developments related to World Wide Web standards may be followed in the W3C Web site (see Section 1.5).

1.3.2 Multimedia

Multimedia can be seen from two perspectives: a variety of analogue and digital data types that come together via common channels of communication

(Laurini and Thompson, 1992); or the merging of three industries: computing, communication, and broadcasting (Fuhrt, 1994). The data types that are now possible to use with multimedia tools include (Gibbs and Tsichritzis, 1995; Kemp, 1995):

- text of infinitely variable size and structure;
- still images, like bitmaps and raster images;
- still and animated computer-generated graphics;
- audio, whether synthesised or captured, and replayed sound;
- video or moving frames.

Text is the most widely used type of media. It is the least demanding data type in terms of storage. Most of the network protocols are text-based.

Still images are the second most used data type in multimedia applications. Their storage requirements are more significant than those of text. They depend on image size, resolution, and colour depth. Compression techniques play a significant role.

Audio is increasingly common in commercial applications. There are a number of audio file formats. Storage requirements are high which makes relevant the use of compression.

Video is the most demanding multimedia data type concerning storage. The video objects are stored as sequences of frames; depending on resolution and size, a single frame can consume more than one megabyte of storage. Realistic video playback requires a transfer rate in the order of 30 frames per second in the NTSC system used in North America and Japan. The European PAL system has a transfer rate of 24 frames per second.

Multimedia software includes electronic games, hypermedia browsers, and authoring and desktop conferencing systems. These applications can be stand-alone or distributed. Distributed applications present advantages such as the use of live cameras for real-time information update (Meleis, 1996). An early application of multimedia with environmental implications was the Massachusetts Institute of Technology (MIT) Aspen Movie Map developed in the 1970s. This Map enabled a virtual exploration of Aspen, Colorado by using film shots that were accessed interactively to simulate driving through the town.

Other pioneers in the use of multimedia for environmental applications have been the Domesday effort described in Rhind *et al.* (1988), where one could have access to videos by clicking on a UK map, and Vining and Orland (1989) and Bishop and Hull (1991) proposals for landscape analysis. Joffe and Wright's (1989) description of SimCity is another early reference. An early application of multimedia spatial information systems was presented by Ertl *et al.* (1992) for urban planning. More recent environmental multimedia systems were described by Graf *et al.* (1994), Groom and Kemp (1995), Blat *et al.* (1995), Fonseca *et al.* (1995), Raper and Livinstone (1995), Shiffer (1995*ab*), Raper (1997), Camara and Raper (1999), and Hu (1999). These applications added multimedia capabilities to geographical information systems. Many are related to the application of multimedia spatial information systems to environmental impact assessment. Several environmental multimedia Web sites are mentioned in Section 1.6.

For general information on multimedia, the interested reader may follow the activities of the Special Interest Group (SIG) on Multimedia from the Association of Computing Machinery and the on-line presence of the Institute of Electrical and Electronics Engineers' *IEEE Multimedia Journal* (see Section 1.6).

1.3.3 Virtual Reality

Virtual reality (VR) enables one 'to get into a screen' and interact with an environment including three-dimensional images and sounds generated in real time. These environments may represent real or abstract spaces. Virtual environments are populated by objects that may have a predefined behaviour and interaction rules. The user can move independently of these objects and interact with them in real time, using his/her senses.

The senses most investigated in VR have been vision, audition, and haptics (Pimentel and Teixeira, 1993; Burdea and Coiffet, 1994; and Stuart, 1996). Olfactory and gustatory displays are not common. Head-mounted displays (HMDs) are used to create visual and auditory displays. They involve the tracking of the user's head position and orientation, giving the computer the parameters to render images and sound in real time. Augmented reality, the superimposition of generated images on real-world views by using see-through HMDs is an application of these devices relevant for environmental problems. Gloves presented as virtual hands, or just real hands associated with a computer vision system have been used to provide visual feedback. There are now devices capable of providing force and tactile feedback, texture, and temperature displays but their commercial availability is limited.

Non-immersive solutions (those not requiring HMDs) are becoming common due to the development of the Virtual Reality Modeling Language (VRML) for the Internet (Ames *et al.*, 1997). They will be discussed extensively in this book. In addition, the Internet will increasingly become a basis for telepresence applications such as NASA's exploration of Mars. Telepresence relies upon the use of robots and live cameras that can be manipulated remotely while images are being transmitted. Environmental applications of VR technologies are described by Wheless *et al.* (1996), Gaither *et al.* (1997), and Camara *et al.* (1998), among others.

A general site on virtual reality may be found at the Human Interface Technology Laboratory at the University of Washington site. The repository on VRML information may be accessed at the Web3D consortium site (see Section 1.6).

Fig. 1.1 Screen shot of Microsoft's TerraServer

Source: http://terraserver.microsoft.com. Image courtesy of Microsoft Research

1.3.4 Environmental Multidimensional Web Sites

To capture the notion of environmental multidimensional web sites, the reader should browse the following sites with addresses listed in Section 1.6. Figures 1.1–1.5 provide some illustrative screen shots of some of the sites.

- *Database-backed Web sites.* Most of the environmental information sites are backed by databases. The largest database backed site in the Internet is TerraServer from Microsoft (Fig. 1.1) with more than 1 terabyte of image data and 4.1 terabytes of uncompressed data. TerraServer relies on US Geological Survey aerial photographs and space information satellite imagery with 2 metres of resolution (SPIN-2).
- *Environmental information infrastructures.* These infrastructures enable the querying of distributed databases by using meta-information on data, models, and tools. There are several environmental information infrastructures already on the WWW such as the Master Environmental Library (MEL) maintained by the US Navy (Fig. 1.2), NASA's Global Change Master Directory, EPA, and Europe's Centre for Earth Observation (CEO).

MEL includes environmental data and modelling, and visualisation tools. The Global Change Master Directory provides access to datasets and other relevant global climate change related information including scholarly electronic journals.

The EPA infrastructure allows the query of databases on drinking water, water and air pollution, toxic releases, and hazardous wastes in the United States. Finally, the CEO system provides access to a variety of information sources related to earth observation. Its collection of case studies on the use of satellite imagery for environmental applications is particularly useful.

There are several infrastructures that include chatrooms. Environonline hosts chatrooms for environmental professionals. Chat topics include water pollution, public works, and solid waste. Earthsystems, which provides access to the Virtual Library on the Environment, offers chatrooms for a variety of environmental science related topics.

- *Environmental modelling and interactive visualisation tool sites.* An illustrative example is the collection of operations research oriented components implemented in Java, available at Michael Trick's Web site at Carnegie Mellon University. There are a large number of other sites enabling the download of environmental software listed in Chapters 4 and 5.
- *Interactive movies.* Videos on interactive environmental applications may be seen at the University of Toronto maintained site (Fig. 1.3) and the

Fig. 1.2 Screen shot of the Master Environmental Library

Source: http://mel.dmso.mil/. The Master Environmental Library (MEL) is a US Department of Defense effort sponsored by the Defense Modeling and Simulation Office (DMSO) (http://www.dmso.mil)

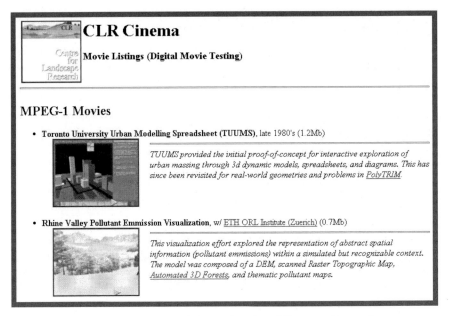

Fig. 1.3 Screen shot of the University of Toronto Center for
Landscape Research (CLR) Web site

Source: http://www.clr.utoronto.ca/movies.html. Image courtesy of
Centre for Landscape Research, University of Toronto

MIT's Monumental Core Virtual Streetscape Web page (Fig. 1.4). This
site includes QuickTime VR clips to explore areas in Washington, DC.
Secrets@sea is an animation oriented Web site for environmental education
purposes (Fig. 1.5).

1.4 Overview of the Book's Web Site

The book's Web site includes an overview of the book, demonstrations, and a
collection of resources. These resources are regularly updated (http://gasa.dcea.
fct.unl.pt/camara/index.html). Also, there is a list of suggested projects and links
to illustrative examples, and a discussion forum.

1.5 Overview of the Book

In addition to this introductory chapter, this book has seven chapters and four
appendices, outlined as follows:

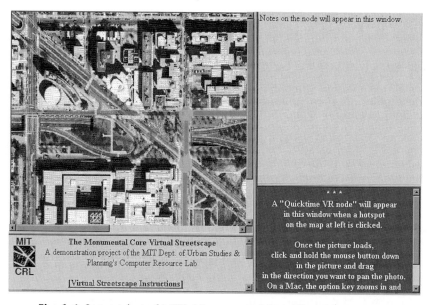

Fig. 1.4 Screen shot of MIT's Monumental Core Virtual Streetscape

Source: http://yerkes.mit.edu/ncpc96/home.html. Image courtesy of Computer Resource Laboratory, Massachusetts Institute of Technology

Fig. 1.5 Screen shot of the Secrets @ Sea Web site

Source: http://www.secretsatsea.org. Image courtesy of Engaging Science

Chapter 2: Multidimensional Environmental Data   The application of remote sensors with high spatial and spectral resolutions, complemented by fibre–optic, laser, electrochemical, piezoelectric, and biosensors in environmental sampling is discussed. Emphasis is placed on multidimensional data types such as static and dynamic images and sound. Two examples included are the application of remote sensing data in desertification and digital video for air pollution analysis.

Chapter 3: Multidimensional Environmental Information Systems The backing of Web sites with databases, geographical information systems, and multimedia databases is reviewed. The development of modern information infrastructures, where several distributed databases can be queried, is discussed. The Expo98 Environmental Information System, the database backed Working Site for the European Spatial Metadata Infrastructure (ESMI), inovaGIS (an open source geographical information system based on components for the Internet), and the National System for Geographic Information of Portugal are illustrative examples.

Chapter 4: Multidimensional Environmental Modelling   This chapter focuses on the application of static and dynamic images in environmental simulation models. Types of data considered include orthophotos, digital terrain models draped with photo–textures, and three-dimensional representations of natural and artificial objects. The use of cellular automata techniques and other individual-based techniques in pollution and ecological modelling are presented. The derivation of simulation models from digital video based on a new technique called Programming by Reproduction is also introduced. Finally, the development of truly multidimensional models integrating numerical, pictorial, and linguistic variables is proposed. Forest fire, water quality, and predator–prey models are illustrative examples included.

Chapter 5: Sensorial Exploration of Environmental Information   This chapter reviews basic concepts and environmental applications of visualisation methods for:

- one- two-, and three-dimensional data;
- uncertainty analysis;
- virtual environments.

Animation techniques used in the visualisation of data are also reviewed. Finally, the exploration of environmental data using other senses, most notably hearing, are briefly introduced. Illustrative example include applications developed for Expo98, the World Exhibition held in Lisbon, Portugal.

Chapter 6: Interaction Design for Environmental Applications   Methods that facilitate the interaction of users with environmental applications are discussed. Emphasis is on existing methods such as sketching, the use of sound, virtual environments, and collaborative tools. Proposals for the use of agents (soft robots) are briefly discussed. Examples cover environmental impact and environmental education applications.

Chapter 7: Towards Interactive Environmental Movies   This chapter reviews initiatives towards the development of interactive movies based on two examples: The Environmental Exploratory System developed for Expo98 and Virtual Tejo, a virtual reality-based water quality management system developed for a stretch of the Tejo Estuary.

Chapter 8: Future Developments   This chapter attempts to present scenarios considering the expected environmental information available, bandwidth, and processing power during the eary 2000s. It also includes a discussion on the potential impacts of alternative computing models on the development of applications for professionals and the general public. Finally, it discusses how universities should prepare future environmental professionals, with reference to these scenarios.

Appendix 1: World Wide Web Tools and Technologies   In this appendix, a synthesis of the collection of resources available on the WWW to facilitate the design, implementation, and maintenance of an environmental multidimensional Web site is included. It also presents methods to advertise a Web site and track its usage.

Appendix 2: Image Processing Primer   This includes a brief review of image processing fundamentals. It also reviews Fourier transforms and wavelet functions, tools that are relevant for topics introduced in Chapters 2, 3, and 5.

Appendix 3: Computer Graphics Primer   This presents a brief introduction to computer graphics concepts required for understanding the visualisation and animation methods presented in Chapter 5.

Appendix 4: Web Addresses of Periodicals   This provides a list of Web addresses of periodicals on the materials covered in this book. Many of these Web sites include abstracts; others provide full text for a paid subscription.

1.6 Further Exploration

The Web provides many environmentally related resources that are sometimes ephemeral. To update the links mentioned below, this book's Web site can be accessed at (http://gasa.dcea.fct.unl.pt/camara/index.html).

Environmental Problems

Digital libraries with environmental bibliography and multidimensional data sources are a relevant development of the World Wide Web. Useful examples are:

- World Wide Web Virtual Library on the Environment (http://earthsystems. org/Environment.shtml). This is a comprehensive site covering resources in ecology, earth sciences, energy, environmental law, forestry, landscape planning, meteorology, oceanography, and sustainable development.
- Environmental Route Net is an environmental digital library maintained by Cambridge Scientific Abstracts (http://www.csa.com/routenet/newaccess. html).
- Envirolink (http://www.envirolink.org) provides access to environmental information sources for the general public.
- Alexandria (http://alexandria.sdc.ucsb.edu). This digital library research effort is directed towards the uses of spatial data in the World Wide Web.
- Berkeley (http://elib.cs.berkeley.edu). Methods to handle photographs, satellite images, tabular datasets, and full text documents in environmental digital libraries are illustrated.

Other digital library efforts that are relevant for environmental applications are based at University of Illinois (http://dli.grainger.uiuc.edu) and Stanford University (http://digilib.stanford.edu).

A wealth of environmental information may also be accessed through institutional sites such as:

- United Nations Environment Program (http://www.unep.org). This site provides access to global environmental information. It includes a directory of Ministries of the Environment around the world.
- European Environment Agency (http://www.eea.eu.int). This site provides information on environmental programs across the European Union.
- United States Environmental Protection Agency (http://www.epa.gov). This Web site is a coordinating infrastructure for EPA's data, methods, and tools.
- United Kingdom (http://www.environment-agency.gov.uk).
- Environment Canada (http://www.ec.gc.ca) and Australia Environment On-Line (http://www.erin.gov.au) are other representative institutional sites.

Environmental problems with spatial dimensions are addressed in Web sites on planning and the environment, environmental impact statements, and conservation, such as:

- United States Geological Survey (USGS) site on data sources for environmental planning and impact studies (http://h2o.usgs.gov/public/eap/env_data.html).
- International Association for Impact Assessment (IAIA) resources for environmental impact analysis (http://www.iaia.org.eialist.html).
- Oak Ridge National Laboratory resources for risk analysis (http://www.hsrd.ornl.gov/ecorisk/ecorisk.html).

Four excellent Web sites on conservation issues are World Wildlife Fund Global Network (http://www.panda.org), Earthwatch International (http://www.earthwatch.com), The Nature Conservancy (http://www.tnc.org), and Conservation International (http://www.conservation.org) Web sites.

Industrial ecology applications may be found at:

- EPA's environmental prevention Web site (http://es.epa.gov).
- MIT's Technology, Business and Environment Web site which includes a bibliography on industrial ecology and a gallery of ecologically designed products (http://tbe.mit.edu/gallery/).
- Product Ecology Consultants site showing methods for ecological design (http://www.pre.nl).

Multimedia and Virtual Reality

Multimedia and virtual reality resources may be found at:

- World Wide Web Consortium, which offers guidelines and standards for the World Wide Web (http://w3c.bilkent.edu.tr).
- ACM Special Interest Group on Multimedia (http://www.acm.org), and the on-line presence of IEEE Multimedia (http://computer.org).
- Simon Gibbs and Gabor Szentivanyi Web site on multimedia resources is no longer maintained but it still is a comprehensive resource on multimedia (http://viswiz.gmd.de/MultimediaInfo).
- University of Washington site is an excellent source on virtual reality research and applications (http://www.hitl.washington.edu).
- The repository on VRML information may be accessed at the Web3D consortium (http://www.web3d.org).

Environmental Multidimensional Web Sites

Illustrative environmental multidimensional Web sites include:

- TerraServer from Microsoft (http://terraserver.microsoft.com).
- Master Environmental Library (http://mel.dmso.mil).
- NASA's Global Change Master Directory (http://gcmd.gsfc.nasa.gov).

- EPA warehouse on environmental facts (http://www.epa.gov/enviro/index.html).
- Europe's Centre for Earth Observation (CEO) case studies on remote sensing applications (http://infeo.ceo.org/).
- Environonline and Earthsystems host chatrooms for environmental professionals (http://www.pollutiononline.com and http://www.earthsystems.org), respectively.
- Operations research software components available at (http://mat.gsia.cmu.edu//program.html) may be used for environmental applications.
- Videos on interactive environmental applications may be seen at the University of Toronto maintained site (http://www.clr.utoronto.ca).
- MIT's the Monumental Core Virtual Streetscape (http://yerkes.mit.edu/ncpc96/home.html) includes QuickTime VR clips to explore corners in Washington DC.
- MIT's Virtual Tour of the Campus (http://web.mit.edu) also uses Quick-Time VR clips.
- A Web based cartoon animation for environmental (marine) education is shown at (http://www.secretsatsea.org).

References

Ames, A.L., Nadeau, D.R., and Moreland, J.L. (1997). *VRML 2.0 Sourcebook*. New York: Wiley.

Berners-Lee, T. (1996). WWW: Past, Present and Future. *IEEE Computer*, 29/10: 69–77.

Bishop, I. and Hull, R.B. (1991). Integrating Technologies for Visual Resource Management. *Journal of Environmental Management*, 32: 295–312.

Blat, J., Delgado, A., Ruiz, M., and Segui, J.M. (1995). Designing Multimedia GIS for Territorial Planning: The ParcBIT Case. *Environment and Planning B*, 22: 665–78.

Burdea, G. and Coiffet, P. (1994). *Virtual Reality Technology*. New York: Wiley.

Camara, A.S., Neves, J.N., Muchaxo, J., Fernandes, J.P., Sousa, I., Nobre, E., *et al.* (1998). Virtual Environments and Water Quality Management. *Journal of Infrastructure Systems, ASCE*, 4/1: 28–36.

Camara, A.S. and Raper, J. (eds) (1999). *Spatial Multimedia and Virtual Reality*. London: Taylor & Francis.

Downs, A. (1972). Up and Down with Ecology: the 'Issue-Attention' Cycle. *The Public Interest*, 6: 38–50.

Ertl, G., Gleixner, G., and Ranziger, M. (1992). Move-X: A System for Combining Video Films, Computer Animations and GIS Data. *Proceedings of the 15th Urban Data Management*, Lyon, 247–54.

Fiksel, J. (1996). *Design for Environment: Creating Eco-Efficient Products and Processes*. New York: McGraw-Hill.

Fonseca, A., Gouveia, C., Camara, A.S., and Silva, J.P. (1995). Environmental Impact Assessment Using Multimedia Information Systems. *Environment and Planning B*, 22: 637–48.

Fuhrt, B. (1994). Multimedia Systems: An Overview. *IEEE Multimedia*, 1/1: 12–24.

Gaither, K., Moorehead, R., Nations, S., and Fox, D. (1997). Visualizing Ocean Circulation Models Through Virtual Environments. *IEEE Computer Graphics and Applications*, 17/1: 16–19.

Gibbs, S. and Tsichritzis, D. (1995). *Multimedia Programming: Objects, Environments and Frameworks*. New York: ACM Press.

Goudie, A. (1986). *The Human Impact on the Natural Environment*. Oxford: Blackwell.

Graf, K.C., Suter, R.M., Hagger, J., and Nuesch, D. (1994). Computer Graphics and Remote Sensing—A Synthesis for Environmental Planning and Civil Engineering. *Computer Graphics Forum*, 13/3: C13–22.

Groom, J. and Kemp, Z. (1995). Generic Multimedia Facilities in Geographical Information Systems. In P. Fisher (ed.), *Innovations in GIS2*, 359–68. London: Taylor & Francis.

Harmancioglou, N.B., Singh, V.P., and Alpaslan, M.N. (eds) (1998). *Environmental Data Management*. Dordrecht: Kluwer.

Hu, S.F. (1999). Integrated Multimedia Approach to the Utilization of an Everglades Vegetation Database. *Photogrammetric Engineering and Remote Sensing*, 65/2: 193–8.

Joffe, B. and W. Wright, W. (1989). SIMCITY: Thematic Mapping + City Management Simulation. *Proceedings of GIS/LIS 89*, 591–600. Falls Church, VA.

Kemp, Z. (1995). Multimedia and Spatial Information Systems. *IEEE Multimedia*, 271: 68–76.

Laurini, R. and Thompson, D. (1992). *Fundamentals of Spatial Information Systems*. London: Academic Press.

Lein, J.K. (1997). *Environmental Decision-making*. Malden, MA: Blackwell.

Loucks, D.P. and da Costa, J.R. (eds) (1991). *Decision Support Systems: Water Resources Planning*. Berlin: Springer Verlag.

Meleis, H. (1996). Toward the Information Network. *IEEE Computer*, 29/10: 59–68.

Newson, M. (ed.) (1992). *Managing the Human Impact on the Natural Environment*. London: Belhaven Press.

Nielsen, J. (1994). *Multimedia and Hypertext: The Internet and Beyond*. Cambridge, MA: Academic Press.

Pimentel, K. and Teixeira, K. (1993). *Virtual Reality*. New York: McGraw-Hill.

Raper, J. and Livingstone, D. (1995). The Development of Spatial Data Explorer within an Environmental Hyperdocument. *Environment and Planning B*, 22: 679–87.

Raper, J. (1997). Progress in Spatial Multimedia. In M. Craggier and H. Couclelis (eds), *Geographic Information Research*, 525–43. London: Taylor & Francis.

Rhind, D.W., Armstrong, P., and Openshaw, S. (1988). The Domesday Machine: A Nation Wide GIS. *Geographic Journal*, 154: 56–68.

Richards, D.J. (ed.) (1997). *The Industrial Green Game*. Washington, DC: National Academy Press.

Shiffer, M. (1995a). Environmental Review with Hypermedia Systems. *Environmental and Planning B*, 22: 359–72.

Shiffer, M. (1995*b*). Interactive Multimedia Planning Support: Moving from Stand-Alone Systems to the World Wide Web. *Environment and Planning B*, 22: 649–64.

Stuart, R. (1996). *The Design of Virtual Environments*. New York: McGraw-Hill.

Vining, J. and Orland, B. (1989). The Video Advantage: A Comparison of Two Environmental Representation Techniques. *Journal of Environmental Management*, 29: 275–83.

Watkins, D.W. and McKinney, D.C. (1995). Recent Developments Associated with Decision Support Systems in Water Resources: 2. *Review of Geophysics, 33(suppl. S)*: 941–8.

Wheless, G.H., Lascara, C., Valle–Levinson, A., Brutzman, D.P., Sherman, W., Hibbard, W., *et al.* (1996). Virtual Cheasapeake Bay: Interacting with a Coupled Physical-Biological Model. *IEEE Computer Graphics and Applications*, 16/4: 52–7.

2

Multidimensional Environmental Data

2.1 Introduction

Lancia, the Italian automobile manufacturer, had a television commercial a few years ago in which a car was travelling on the top of a map draping a digital terrain model of Italy. Now and then, the map would morph into a video of the natural landscape. The symbolic and the realistic were fully intertwined (Lancia, 1993).

In the environmental professions, realistic images using photography (both aerial and ground-based) have been used for decades. But data analysis has relied mainly on symbolic representations such as maps and alphanumerical data. Multimedia and new sensor technologies bring together realistic and symbolic views of environmental phenomena.

Multimedia and sensor technologies, and the growth of the Internet are certainly changing the way environmental agencies acquire, handle, and make available environmental data. This chapter discusses the acquisition stage and, in particular, the collection and use of image and sound data from remote and local sensors.

Environmental data is needed in order to understand the effects of human actions and natural factors on environmental processes. It is also required to determine the effectiveness of pollution control measures. In addition, data are necessary inputs for environmental impact assessments, urban and regional plans, and modelling efforts. The periodicity of data collection programs depends on the nature of environmental attributes. These can be divided into

Table 2.1 Examples of environmental attributes

'Permanent' attributes
  Topography
  Land use
  Water bodies

'Semi-permanent' attributes
  Vegetation (cover, type, plant community composition)
  Distribution of non-migratory fauna

Ephemeral or seasonal attributes
  Water quality
  Air quality
  Sound (noise and natural sounds)
  Weather conditions
  Soil properties (i.e., moisture)
  Vegetation (greenness, plant productivity)
  Distribution of migratory fauna

Source: Adapted from Clarke, 1986.

'permanent', 'semi-permanent', and ephemeral or seasonal attributes (Clarke, 1986), as illustrated in Table 2.1.

Environmental data is collected at global, international, national, regional, and local scales. The monitoring programmes of the United Nations Environmental Program (UNEP) and the International Geosphere–Biosphere Program (IGBP) are examples of international initiatives. Other representative efforts include the directory of global change data that can be found at NASA's Global Change Master Directory and the European Environment Agency's infrastructure for European environmental data (see Chapter 1 and Section 2.7). There are also several information sources covering disasters, such as floods, fires, volcanic eruptions, earthquakes, and oil spills, with an international scope (Section 2.7). Most countries have national monitoring programmes that are supplemented by data collected in regional and local studies. Table 2.2 shows the United Kingdom's environmental monitoring register as an illustrative case.

There are usually several agencies in the same country collecting environmental data (typically, mapping, soil surveys, environmental quality, and natural resources). Even in the same agency there are different departments that work independently. A significant advantage introduced by the World Wide Web is that, for the first time, national information infrastructures integrate data that were traditionally dispersed in agencies and departments.

A comprehensive example is provided by the United States Environmental Protection Agency (USEPA or simply EPA) and the Environment Defense

Table 2.2 The United Kingdom's environmental monitoring register

---

Air: emissions of pollutants from factories
    Urban and rural networks for smoke and sulphur
        dioxide
    Sulphur dioxide baseline network
    Nitrogen oxide networks
    Particulate monitoring schemes
    Lead in air
    Acid deposition network

Fresh water
    Industrial discharges
    Surface and groundwater quality
    Discharges from wastewater treatment plants

Sea water
    Discharges into estuaries and coastal waters
    Bathing water quality
    Estuaries water quality
    Water disposal at sea
    Chemical and biological monitoring of the sea

Land
    Use
    Disposal of wastes
    Quality

Waste
    Disposal sites
    Special and hazardous wastes
    Recycling
    Leaching

Fauna and flora
    Habitat quality
    Endangered and extinct species
    Health
    Radioactivity and heavy metal levels

Source: Adapted from Newson, 1992.

Fund which enables any member of the public to access environmental data available for any location in the United States, querying by zip code (see the Scorecard system, Section 3.2.4). Another interesting EPA programme involved with public participation in monitoring is described on the site referenced in

Section 2.7. The EPA site also provides access to Web sites of similar agencies around the world. Examples of well-developed national programmes may be found in the Web pages for Australia and Canada (see Chapter 1).

The rise of multimedia technologies and the Internet provides other significant advantages for environmental monitoring programmes: the improved capabilities for handling images (both still and video) and sound data, and integrating different data types (images, sound, text). The Internet is also an effective medium for allowing environmental data to be available to the public.

In this chapter, the nature of multimedia data and their role in valuing 'permanent', 'semi-permanent', and ephemeral environmental attributes will be examined. The chapter is divided into five sections:

- *Remote sensing imagery*, focusing on the use of high spatial and spectral resolution images obtained by satellites and aircraft. This section includes notes on image processing. The complementary application of a global positioning system (GPS) is also reviewed.
- *Photography*, discussing aerial photography and the use of ground level photos.
- *Videography*, covering the use of aircraft-based as well as ground video-recording devices.
- *Sound*, discussing both unwanted sound (noise) and natural sounds.
- *Other sensors*, briefly examining the role of fibre-optic, laser, electro-chemical, electronic noses, and biosensor devices. A reference to human sensing is also included.

The use of sensing devices, particularly those with scanning capabilities, such as remote sensors associated with wide public collaboration, may radically change the nature of environmental monitoring. Consider the daily availability of satellite images with less than one-metre resolution and laser sensors scanning air pollution levels over wide areas. Think also of a large number of households in a city who could analyse their drinking water, measure noise with sensors, and report the results to a Web page.

Today, this stage has not yet been reached. The traditional selection of sampling sites, frequencies, variables, and sampling duration remains necessary. The definition of the type of sampling (random, systematic, restricted random, stratified, or selective) is also an essential step. These concepts are not reviewed here but can be found in classic texts, such as ACS (1988) and Gilbert (1987) and in papers, such as Ward and Loftis (1986), Whitfield (1988), Moss (1989), Loftis *et al.* (1991), Trujillo-Ventura and Ellis (1991). More recent proposals may be found in Maher *et al.* (1994), Cox and Piegorsch (1996), and in the collection edited by Harmancioglou *et al.* (1998).

2.2 Remote Sensing Imagery

2.2.1 Nature of Remote Sensing

Schott (1997) defines remote sensing as the field of study associated with capturing information about an object without coming into physical contact with it. Remote sensing may include many areas ranging from astronomy to medical imaging. In this book, remote sensing is restricted to the use of spacecraft (and aircraft) remote sensing imagery for earth observation (EO) purposes. The use of remote sensing images for environmental applications is well documented in the collection of case studies provided in the Centre for Earth Observation (CEO) Web site (Section 2.7). The quantity, quality, and range of such applications are increasing considerably with the emergence of high-resolution satellite images.

To maintain reasonable spatial resolutions, most EO systems use low earth orbits (LEOs). These orbits are commonly circular, to facilitate comparable image scales anywhere on the earth (Schott, 1997). To avoid correcting for illumination angle variations in multiple overpasses, sun synchronous orbits are often selected. In these orbits, the satellite crosses the equator (and over points north and south) at approximately the same solar time with each pass. Another type of circular orbit is the geosynchronous earth orbit (GEO), which has a 24 hour period. In GEO, the satellite velocity compensates the earth's rotation rate and the satellite becomes located over a fixed longitude.

Remote sensing images resulting from EO systems are grids (or raster) of cells (or pixels) that can be defined by the following characteristics (Jensen, 1996; Molenaar, 1998):

- *Spatial resolution*, which is a measure of the smallest angular or linear separation between two objects obtained by the sensor.
- *Spectral resolution*, that refers to the number of observed bands of the electromagnetic spectrum to which the sensor is sensitive and the width of each band in terms of frequency or wavelength.
- *Radiometric resolution* specifying the radiance or intensity levels observed for each spectral band.
- *Temporal resolution*, which refers to how often the sensor records imagery of a particular area and is related to the sensor's ground field of view or swath width, a measure of its coverage capability.

Remote sensing systems can be classified as passive or active. Passive systems register naturally occurring radiation that is reflected or emitted from the terrain in the visible, infrared, and other regions of the electromagnetic spectrum (Fig. 2.1). Active systems, such as microwave (radar) and sonar, emit electromagnetic

LANDSAT Bands

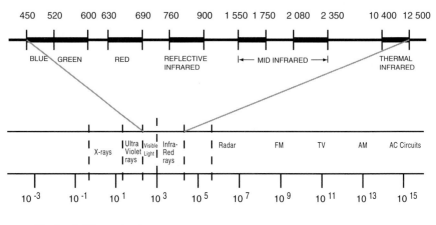

Wavelength (Nanometers)

Fig. 2.1 The electromagnetic spectrum

energy and then measure and time the backscatter radiation returned to the sensor system.

Vincent (1997) argues that extracting environmental information from remote sensing is similar to the work of a spectroscopist: the key is to be able to interpret rapid changes of spectral reflectance, absorption, and emittance as a function of wavelength. The goal is to distinguish between signals that are spatially near or spectrally similar. The bands are normally selected to enhance the contrast between the object of interest and its background (Jensen, 1996). This is possible with multispectral systems, such as the LANDSAT Thematic Mapper, where the sensors record specific ranges of the radiant flux. By contrast, panchromatic black-and-white aerial photography operates in a wide band. In this case, the sensor records all the reflected blue, green, and red radiant flux. Multispectral systems operate over a discrete set of bands. The new hyperspectral systems provide near continuous radiometry over the visible, near infrared to short-wave infrared spectrum.

Earth observation sensors (both space and aircraft systems), can be divided into three major groups (see also Table 2.3 and the image gallery in Fig. 2.2):

- multispectral and hyperspectral sensors;
- high spatial resolution sensors;
- RADAR sensors.

Table 2.3 Selected sensors for the year 2001

Sensor type	Program	Panchromatic resolution (m)	Multispectral resolution (m)	Radar resolution (m)	No. of colour bands	Temporal resolution
Multispectral	GOES		1000		5	0.5 hours
	AVHRR		1100		5	14.5 days
	SeaWiFS		1000		8	1 day
	LANDSAT 7	15	30		7	16 days
	SPOT 5	10	20		4	Pointable
Spatial high resolution	Quick Bird	1	4		4	2–3 days
	SPIN-2	2				2–3 days
	IKONOS	1	4		4	2–3 days
Radar	Radarsat			Variable		1–6 days
	ERS-1			30		Variable

Source: Adapted from Jensen, 1996; Carlson and Patel, 1997.

Jensen (1996), ASPRS (1996), and the Canadian Center for Remote Sensing provide extensive reviews. For a directory of earth observation sensors see the NASA Web site (Section 2.7).

Multispectral and Hyperspectral Sensors

Sensors that offer broad-area coverage have 5–30 m spatial resolution and are multispectral or hyperspectral. Multispectral solutions include sensors installed in satellites for weather, marine, and land observation. For weather observation, Geostationary Operational Environmental Satellite (GOES) and the Advanced Very High Resolution Radiometer (AVHRR) are two illustrative programmes.

GOES was designed to provide the US National Weather Service with frequent imaging of earth's surface and cloud cover. GOES satellites have five channels sensing visible and infrared reflected and emitted solar radiation, and observe the earth continuously at a nadir pixel resolution of 4 km. They also have flexible scanning capabilities. The entire satellite, located at a geosynchronous orbit, rotates to provide a line scan effect, and a mirror oscillates to produce the line advance (Schott, 1997). AVHRR, aboard the NOAA satellites, detects radiation in the visible, near, and mid infrared and parts of the thermal infrared spectrum. Depending on the bands, AVHRR provides cloud, snow, and ice monitoring, sea surface temperature, volcano, forest fire activity, and soil moisture estimates.

SeaWiFS, with a spatial resolution of 1.13 km, is an advanced sensor for ocean monitoring which is aboard the SeaStar spacecraft. It has eight spectral bands that aid monitoring ocean primary production and phytoplankton processes, ocean influences on the climate and the carbon, sulphur, and nitrogen cycles. A

Fig. 2.2

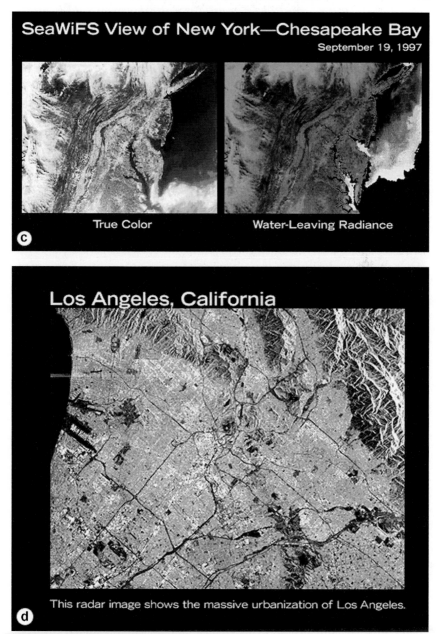

Fig. 2.2 Gallery of satellite images: (a) Place de la Nation, Paris (SPIN-2); (b) Amsterdam (LANDSAT); (c) New York, Chesapeake Bay (SeaWifS); (d) Los Angeles (radar image)

Sources: Fig. 2.2a: Image courtesy of SPIN-2; Fig. 2.2b: NASA Observatorium (http://observe.ivv.nasa.gov/nasa/education/tools/stepby/multi.html); Figs. 2.2c and 2.2d: NASA's Destination Earth project (http://www.earth.nasa.gov)

scanner tilt mechanism enables the instrument to be oriented to avoid sun glint from the sea surface.

The United States' LANDSAT, French SPOT, India's IRS, a joint effort by China and Brazil, Resource 21, a Boeing affiliate and the UK XSTAR company programmes are land-oriented. The last two are directed towards agribusiness and want to activate precision farming. They deliver accurate information for managing temporal and spatial variability on farmed land. Their ground swaths vary between 120 km and 200 km. Global coverage times range from 16 days (LANDSAT 7) to 26 days (Chinese–Brazilian programme). The panchromatic resolution varies between 6 m and 20 m. The multispectral resolution is in the 10–30 m range, except for the IRS-P5 with 6 m. All these sensors operate in a number of discrete bands in the spectrum.

LANDSAT is a highly successful programme which was launched in 1972. It has a combination of sensors with spectral bands suitable for earth observation, with acceptable spatial resolution, swath width, and revisiting period. SPOT has a finer spatial resolution and the capability of pointing sensors, allowing it to produce stereoscopic imagery.

Multispectral remotely sensed data can be used to define geographical location and estimate vegetation features (chlorophyll concentration, biomass, foliar water content), surface temperature, soil moisture, evapotranspiration, discrimination of mineral and rock types, and snow and sea ice characteristics (Foody and Curran, 1994; Jensen, 1996). There are many other multispectral systems in place and planned for the near future and can be studied by accessing the general remote sensing sites in Section 2.7.

Hyperspectral solutions were pioneered by CASI, the Compact Airborne Spectrographic Imager that has been used aboard aircraft. It has 288 channels in the visible and infrared wavelengths. The TWR Company offers the HyperSpectral Imager (HSI) with 384 wavelengths. The Airborne Visible InfraRed Imaging Spectrometer (AVIRIS) is another world class instrument in the realm of earth remote sensing. It is a unique optical sensor that delivers calibrated images in 224 contiguous spectral channels. The instrument is aboard a NASA ER-2 aeroplane.

By measuring the continuous spectrum, hyperspectral sensors enhance mineral identification, vegetation management, and stress identification (one can separate healthy from unhealthy ecosystems).

High Spatial Resolution Sensors

High spatial resolution sensors offer moderate spatial and temporal coverage, high spatial resolution (in the order of one metre) and reduced multispectral capabilities. These include sensors that are part of Earth Watch's Quick Bird, Space Imaging's CARTERRA 1 and IKONOS, SPIN-2 satellites, and Orbimage's OrbView3 and OrbView4.

The high spatial resolution of these satellites is achieved at the expense of a moderate spatial coverage (4–36 km ground swath width) and temporal coverage (total coverage in periods ranging from four months to two years). However, these satellites are designed to indicate rapidly any given site and provide images in two to three days. Their panchromatic resolution varies from 0.82 m to 3 m and the multispectral resolution is in the 4–15 m range. The colour bands are used only to provide additional information to panchromatic band data, except for OrbView4, which will include hyperspectral imagery with up to 280 spectral bands.

Images provided by these high-resolution satellite systems will compete with aerial photography. They will be used for traditional mapping applications, environmental impact studies, and real estate. One of the most promising applications is the use of high-resolution images to support media professionals, an initiative named Geo@News (Section 2.7) at the New University of Lisbon.

RADAR Sensors

The RADAR (RAdio Detection And Ranging) sensors come in a variety of resolution/swath combinations. Radar operates in part of the microwave region of the electromagnetic spectrum, specifically over the frequency interval from 40,000 to 300 megahertz (MHz), the later extending just into the higher frequency end of the radio (broadcast) region. Unlike other sensors that passively sense radiation from targets illuminated by the sun or thermal sources, radar generates its own illumination as bursts or pulses of energy are directed to the target and then senses the fractions of the energy returned. By supplying its own illumination, radar systems have the advantage, in contrast to optical systems, of not being limited by atmospheric conditions, such as clouds. They can also function both during day- and night-time. Radar systems provide information on both land (e.g., moisture, texture, vegetation biomass, and foliar water content) and ocean (e.g., ice, surface wind speed and direction, oil spill information) environments.

Examples of imaging radar include Canada's Radarsat and the European Space Agency's ERS-1. An example of a non-imaging radar system is Topex-Poseidon. This system uses radar to measure sea surface height in ice-free oceans. These data have been used to calculate the speed and direction of ocean currents (Jensen, 1996, citing Jones, 1992).

Overall, there will be more than thirty high-resolution satellites plus numerous aircraft carrying multispectral and spectral sensors contributing to earth observation in the next five to ten years. This high number is required to ensure global coverage if satellite imagery is to be a reliable source of spatial information for disaster and relief planning in real or near-real time. There are problems with operational stability (as witnessed by accidents involving LANDSAT 6, Lewis, SPOT 3, and Quick Bird 1), thus requiring this high number of sensors.

Other remote sensing alternatives include photography, videography, and LIDAR (LIght Detection And Ranging) which are discussed below. Additional methods are FLIR (Forward Looking InfraRed) and laser fluoresensors. FLIR sensors are positioned on aircraft or helicopters and provide images ahead of the platform. They are used in search and rescue operations and forest fire monitoring.

Laser fluoresensors are based on the principle that some targets fluoresce, or emit energy, on receiving energy. This occurs due to an absorption of the initial energy, excitation of the molecular components of the target materials, and emission of longer wavelength radiation which is then measured by the sensor (Canadian Center for Remote Sensing, 1998). These sensors are applied for ocean applications to detect oil spills, and map chlorophyll (Gower and Borstad, 1990; Roesler and Perry, 1995).

Remote sensing images have provided immediate benefits in weather forecasting. They have also helped in providing information for short-term events, such as forest fires and oil spills, and long-term phenomena, such as climate change and the Antarctic ozone hole. There are, however, several limitations as discussed by Parkinson (1997):

- Satellite instruments do not collect water, air, or vegetation samples. They receive and transmit radiation values, which are then converted in the desired environmental parameters using algorithms and interpretations. These are sometimes flawed.
- In the visible and near-infrared images, clouds can obscure the surface data. In general, the presence of the atmosphere between the surface and the satellite introduces complications in measuring surface variables for any wavelength.
- The requirement for visible light is a limitation for obtaining visible images in the polar regions, where the polar night lasts four months at a time. As almost all ultraviolet light is from solar radiation, measurement of stratospheric ozone levels in the polar regions is limited for the same reason.
- Some surface variables interfere with others. An example provided by Parkinson (1997) is the effect of snow covering vegetation in measuring the latter.

Parkinson (1997) mentions that earth observation activities will benefit in the early 21st century from two types of improvement:

- Instrument improvements associated with the use of laser technology to measure surface elevation and an increase in the number of channels on infrared sounders. The latter will contribute to more accurate temperature measurement within the atmosphere.

• Developments in algorithms will improve the correction for atmospheric conditions when deriving information on surface variables. Algorithms will be also based more on physical laws than those currently available.

Other significant developments will be related to the rapidly growing field of multisensor data fusion (Varshney, 1997). The theory here is that fusion of complementary information available from different sensors will yield more accurate results than those obtained by a single sensor.

2.2.2 Image Processing

Processing of remotely sensed images is a requirement for correcting distortions and facilitating visual- or machine-based analyses. Remotely sensed images are subject to the following processing stages: rectification of the data to a map projection (pre-processing); image enhancement; image transformation; classification; and change detection. The goal is to summarise remote sensor data as enhanced images, image maps, spatial database files, and statistics or graphs (Jensen, 1996).

Pre-processing

Pre-processing objectives are to correct for sensor- and platform-specific radiometric and geometric distortions of data. Radiometric distortions may be due to variations in scene illumination and viewing geometry, atmospheric conditions, and sensor noise and response. To correct for variations in illumination and geometry between images, one has to model the geometric relationship and distance between the area of the earth's surface imaged, the sun, and the sensor (Canadian Center for Remote Sensing, 1998). To correct for the radiation scattering effects of the atmosphere, a variety of methods, from detailed atmospheric modelling to practical methods, can be used. A simple technique is to subtract the minimum observed value, determined for each specific band (minimum values vary from band to band), from all pixel values in that band. The correction of sensor noise is achieved by calibrating the platform instruments.

Geometric distortions are due to the earth's rotation and curvature, variations in the platform position (altitude, orientation, velocity), and terrain relief. Several of these variations are systematic and can be considered when modelling the sensor and platform motion. Other errors are random and require geometric registration of the image. Registration implies the identification of ground control points in the distorted image and their matching of their true positions as represented in a map (image-to-map registration). Images can also be registered to other images (image-to-image registration). To correct geometrically the original distorted image, it is necessary to determine the values of the new pixel locations of the corrected output image. This is called re-sampling and can be achieved by (Jensen, 1996; Canadian Center for Remote Sensing, 1998):

- *The nearest neighbour method*, where the value of the pixel in the original image which is nearest to the new pixel location is adopted.
- *Bilinear interpolation*, where re-sampling uses a weighted average of four pixels in the original image nearest to the new pixel location.
- *Cubic convolution*, where sixteen pixels of the original image closer to the new pixel are utilised.

Both the bilinear and cubic convolution methods produce sharper images than the nearest neighbour method.

Image Transformation

Image transformations generate new images from two or more sources of image data. These sources may be multiple bands of data from a single multispectral image or from two or more images of the same area taken at different times. Simple transformation methods are matrix operations that include the subtraction of brightness values of pixels of two images and spectral ratioing. Image ratioing can be used to ensure contrast and highlight features (see Box 2.1).

Box 2.1 *Spectral Ratioing*

Seixas (1998) used LANDSAT images to evaluate desertification in southern Portugal. From fieldwork, five thematic classes were identified:

- Dense vegetation, with cork and holm-oaks, pine trees, and a soil cover of scrub vegetation.
- Scrubland, dominated by bushes, shrubs, and herbaceous plants.
- Scarce vegetation, with large areas of bare soil and ephemeral plants.
- Stubble, including wheat wastes from harvest activities.
- Fallow ground.

In August, because of the weak contrast between soil, stubble, dry vegetation, and shadows, there was spectral confusion on the LANDSAT images of the area. However, three band-ratio images were then tried, with success (see Fig. 2.3):

- TM4/TM3, a vegetation index that enabled the discrimination between vegetative (from red to yellow) and non-vegetative area (green);
- TM5/TM7 used to discriminate stubble (cyan);
- TM3/T7 suited to discriminate the non-cultivated fallow ground areas (dark blue).

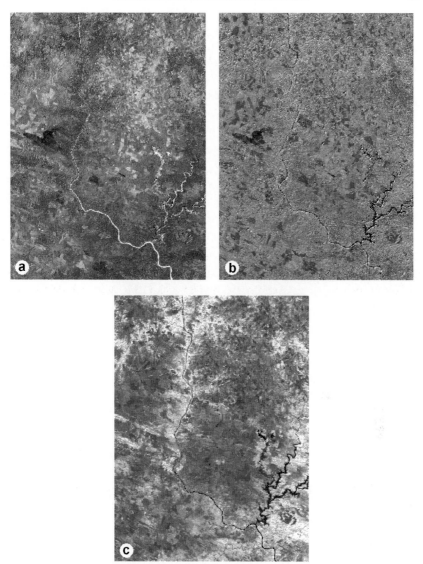

Fig. 2.3 Band-ratio images for vegetation analysis in southern Portugal:
(a) Landsat TM3/TM7; (b) Landsat TM5/TM7; (c) Landsat TM3/TM7
Source: Seixas, 1998

Different bands of multispectral data are usually highly correlated. To reduce data redundancy and correlation between bands, principal components analysis is applied. This technique can be used either to facilitate visual interpretation or reduce the number of bands to be used in image classification.

Fig. 2.4 A red, green, and blue (RGB) composite image used in the analysis of vegetation in southern Portugal
Source: Seixas, 1998

Box 2.1 *Continued*

The red, green, and blue (RGB) false colour composite (Fig. 2.4) with the three ratio images, revealed a meaningful contrast between the fallow (dark blue) and stubble (cyan), the vegetation areas (from red to yellow), and the mixture bare soil and scrub areas (green). A supervised classification was developed, using the maximum likelihood algorithm, being the accuracy coefficients on a range from 86% to 96%.

Image Enhancement

The advantage of digital images is that their pixel values can be manipulated. This manipulation attempts to optimise the images for subsequent visual interpretation. The most widely used image enhancement technique is contrast enhancement. The goal of this technique is to increase the contrast between object targets and their backgrounds. Figure 2.5 illustrates the concept of brightness associated to each pixel location.

The key to contrast enhancement is the concept of an image histogram (Fig. 2.6). This displays brightness values on the x-axis (256 levels) and the frequency of occurrence of these values on the y-axis. By manipulating the histogram and, thus the range of digital values in an image, the contrast and

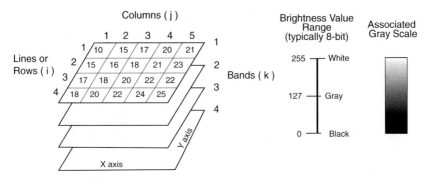

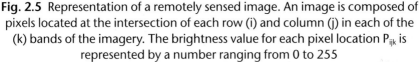

Fig. 2.5 Representation of a remotely sensed image. An image is composed of pixels located at the intersection of each row (i) and column (j) in each of the (k) bands of the imagery. The brightness value for each pixel location P$_{ijk}$ is represented by a number ranging from 0 to 255

From Jensen (1996, fig. 2-1), with permission

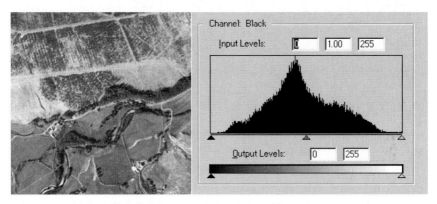

Fig. 2.6 Example of an image histogram

Image from Bower, England, courtesy of SPIN-2

detail in an image can be enhanced. Techniques to achieve this goal include the identification of lower and upper boundaires for brightness values, extension of the boundaries, and linearly stretching all the other values between the two new boundary values. They are illustrated in Fig. 2.7. These techniques are best applied when images have brightness values falling within a single, narrow range of values and one mode (Gaussian) (Jensen, 1996). However, when images include land and water bodies, their histograms become bi-modal or tri-modal, one may apply linear contrast enhancements in a piecewise fashion for each mode (Jensen, 1996).

Histogram equalisation, also shown in Fig. 2.7, is a nonlinear contrast enhancement technique. Its underlying algorithm passes through the individual

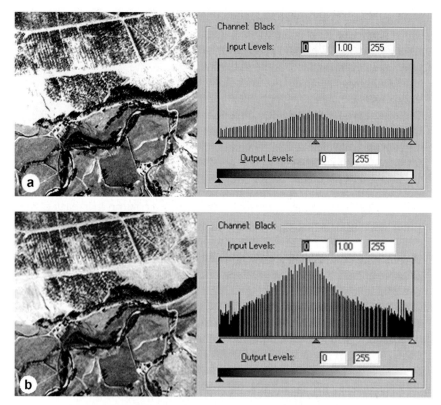

Fig. 2.7 Application of minimum–maximum contrast stretching (a) and histogram equalisation (b) applied to the image from Fig. 2.6

bands of the dataset and assigns a similar number of pixels to each of predefined output grey-scale classes. It applies the greatest contrast enhancement to the most populated range of brightness values.

Other enhancement methods include spatial filtering techniques applied to subsets of an image. They are designed to highlight or suppress image features depending on their spatial frequency. Spatial frequency refers to the number of changes in brightness value per unit distance for any particular subset of the image (Jensen, 1996). It is related to the concept of texture. Spatial filters produce images where their brightness values at any location i,j is a function of some weighted average of brightness values located in the neighbourhood around the i,j location in the input image. The notion of neighbourhood is associated to the one of a 'convolution mask', which can be of different sizes (3×3, 5×5, 7×7) and shapes (square, octagonal, cross). This convolution is operated as a window by moving from pixel to pixel in the image. Figure 2.8 illustrates the application

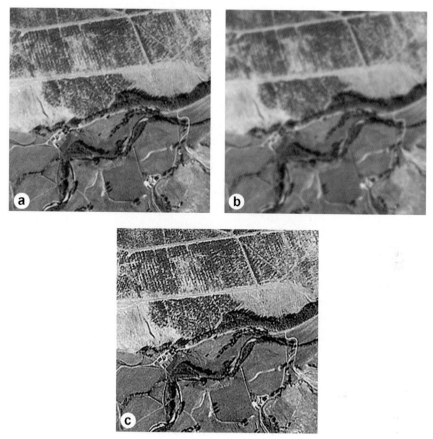

Fig. 2.8 Application of filters to an image (a): blurring (b),
and sharpening filters (c)

of two filters operating with 3×3 convolution masks. Jensen (1996) provides an extensive review on enhancement methods based on spatial filters and their applications in remote sensing. A description of these methods is included in Appendix 2.

Classification

There are two methods for transforming remote sensed images into thematic images (the goal of image classification): supervised and unsupervised. In supervised classification, the analyst begins by identifying, in the image, homogeneous samples of the different classes of crops, forest types or land uses he/she is interested in. The homogeneous samples are named 'training areas'. The different categories are called 'information classes'.

In this identification, he/she takes into account seven classic features in the visual interpretation of an image: tone, shape, size, pattern, shadow, and association (Canadian Center for Remote Sensing, 1998):

- Tone refers to the relative brightness or the colour of objects in an image.
- Shape refers to the form or outline of objects in an image.
- Size of objects is a variable that has absolute and relative dimensions. The size of objects in an image depends on scale.
- Pattern refers to the spatial arrangement of objects in an image.
- Shadow enhances topography and provides an idea on the relative heights of objects. It can also occlude objects.
- Association considers the relationships between nearby objects (i.e., there are roads near a commercial centre).

The selection of the training areas is based on the analyst's familiarity with the real surface cover of those areas. The digital information in all the spectral bands for these areas pixels is then used to train the computer to recognise similar areas for each category, from a spectral standpoint. Thus, information classes are used to determine the spectral categories that represent them (Canadian Center for Remote Sensing, 1998).

In unsupervised classification, the process is reversed. Spectral categories are first grouped using clustering algorithms. The analyst predefines the parameters of these algorithms: number of categories he/she should look for; distance among clusters; and variation within each cluster. The process is iterative, until the analyst is satisfied with the results of the classification process.

Change Detection

Change detection techniques enable the identification of changes in features, such as coastlines, or the alteration of land use. It can also be used to track oil spills. This is achieved by comparing two or more images of the same area. Ideally, change detection analysis should be based on images acquired with a sensor that keeps constant its temporal, spatial (including view angle), spectral, and radiometric resolutions (Jensen, 1996). One should also use images obtained with identical environmental variables such as atmospheric, soil moisture, vegetation, and tidal conditions. There are several algorithms for change detection reviewed by Jensen (1996). Three popular methods are:

- *Post-classification comparison.* This method requires rectification and classification of each remotely sensed image. The classified images are then compared in pairs on a pixel-by-pixel basis using a change detection matrix. A simplified version of this process is shown in Figs 2.9 and 2.10.
- *Multichange detection using a binary change mask.* In this method, one selects a base image that will be compared to another image named 'date 2'. This image is classified using rectified remote sensor data. Then, for a certain

	Vegetation	Water	Burned area	Urban areas	Roads
R	38 - 55	0 - 2	74 - 197	226 - 233	104 - 126
G	58 - 244	0 - 2	19 - 43	146 - 219	84 - 138
B	24 - 35	0 - 2	29 - 90	200 - 226	129 - 151

(a)

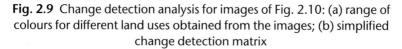

94 \ 95	Burned areas	Urban areas	Roads
Vegetation			
Urban areas		—	—
Roads		—	—

(b)

Fig. 2.9 Change detection analysis for images of Fig. 2.10: (a) range of colours for different land uses obtained from the images; (b) simplified change detection matrix

band, comparing the two images identifies change pixels (methods such as band ratio or subtraction may be applied). The analyst selects a threshold value to identify the areas of change and, as a result, a change pixel mask is created. This mask is overlaid on the date 2 image. Only those pixels that were detected as having changed are classified in the date 2 image. A post-classification comparison can be then applied to provide the from–to change information (Jensen, 1996). Figure 2.11 illustrates the process.

- *Manual on-screen digitisation of change.* This method uses a map, satellite image, or orthophoto as a source. This source is then compared with the date 2 image on the screen side by side.

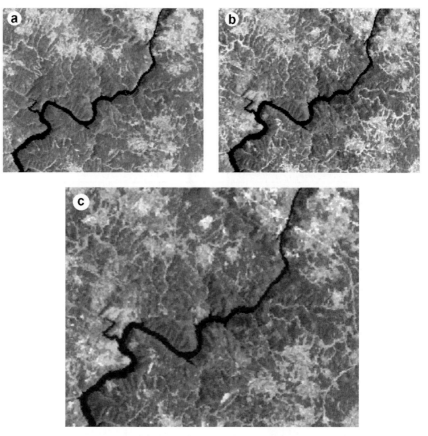

Fig. 2.10 LANDSAT (RGB 743) images of 1994 (a) and 1995 (b) of an area in central Portugal where a fire occurred in 1995. The identification of the burnt vegetation area (c) was made by colour analysis

Images (a) and (b) courtesy of the National Center for Geographic Information, Portugal; and Eurimage, Italy

There are several software packages that may be used for image processing and are listed, together with their Internet addresses, in Section 2.7.

2.2.3 Global Positioning Systems

Global positioning system (GPS) receivers obtain latitude, longitude, and altitude data with reasonable accuracy. They are space-based radio positioning systems that provide 24-hour three-dimensional position (latitude, longitude, altitude), velocity, and time information to suitably equipped users anywhere on or near the surface of the earth. There are currently two 'public' GPS systems:

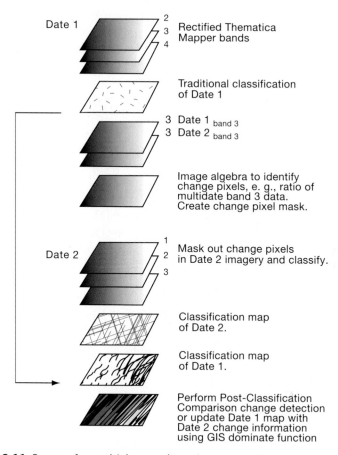

Date 1 — 2 3 4 Rectified Thematica Mapper bands

Traditional classification of Date 1

3 Date 1 band 3
3 Date 2 band 3

Image algebra to identify change pixels, e. g., ratio of multidate band 3 data. Create change pixel mask.

Date 2 — 1 2 3 Mask out change pixels in Date 2 imagery and classify.

Classification map of Date 2.

Classification map of Date 1.

Perform Post-Classification Comparison change detection or update Date 1 map with Date 2 change information using GIS dominate function

Fig. 2.11 Process for multichange detection using a binary change mask

From Jensen (1996, fig. 9-17), with permission

NAVSTAR and GLONASS. The NAVSTAR system is owned by the United States and is managed by the Department of Defense. Since May 2000, GPS precision was improved due to President Clinton's decision to discontinue the use of the principle of selective availability (SA) adopted by the NAVSTAR system. The SA principle limited the accuracy of GPS to within 100 metres. Following Clinton's decision, GPS precision is now within 20 metres. The Russian Federation manages the GLONASS system. The European Union, through the Galileo satellite programme, is also involved in further enhancement of the precision of GPS.

GPS receivers calculate a position by measuring simultaneously the distance of a receiver to three or more satellites that transmit continuous L-band radio signals (Gibbons, 1992). These microwave signals include three pieces of information: satellite identification; precise time provided by atomic clocks on the satellites; and the orbital location of the satellite. Each satellite emits its message with a unique pseudo–random noise code (PRN) that distinguishes its signal from other GPS space vehicles.

By computing distances to satellites using the formula of the speed of light and radio signals multiplied by the time since the signal transmission, a GPS receiver can obtain latitude and longitude coordinates. By using four or more satellite signals, it is possible to obtain altitude. An extra satellite is used to adjust for the differences resulting from the use of a very precise atomic clock in the satellites and less precise clocks in GPS receivers (Gibbons, 1992). GPS receivers have an accuracy of approximately 20 m. Differential methods (called DGGPS) which use a receiver at a precisely known point and transmit these to operating receivers nearby have an accuracy in the order of centimetres (Hurn, 1993).

GPS collars that memorise positions at given points in time are being used in animal tracking; Rodgers *et al.* (1996) and Rutter *et al.* (1997) provide examples. For a discussion on existing alternative animal tracking methods (focusing on livestock), see Frost *et al.* (1997). Summers and Feare (1995) provide an example of the use of radio in tracking birds; the use of electronic tags in fish is discussed by Metcalf and Arnold (1997).

Other applications of GPS data to environmental problems include measuring water vapour and evaluating the movement of the earth's tectonic plates (see Section 2.7 for Web sites on these projects).

2.3 Photography

2.3.1 Aerial Photography

Aerial photography has been used for environmental applications such as environmental impact assessment (e.g., Cohen *et al.*, 1995, Knott *et al.*, 1997) and coastal management (e.g., Gorman *et al.*, 1998). Cohen *et al.* (1995) explored the integration of aerial photos in geographical information systems (GIS) for environmental studies. Knott *et al.* (1997) applied aerial photos in the assessment of the impacts of pipelines on flora and fauna. Aerial photography can use panchromatic black-and-white, colour, or colour infrared film, depending on user goals. Aerial photos can be taken vertically or with an oblique view from manned or unmanned aircraft. Vertical photos are a realistic counterpart to the information provided by a map of an area. Oblique photos, usually obtained at 30 or 60 degrees, provide a three-dimensional effect (Fig. 2.12).

By removing image displacement one can rectify aerial photos and obtain orthophotos. Image displacement is due to terrain relief, plane tilt, and camera

Fig. 2.12 Example of an oblique aerial photo: the Expo98 area in 1993
Source: http://www.expo98.pt/ambiente. Image courtesy of Parque Expo98

lens distortion. Examples of orthophotos available on the World Wide Web for browsing, is MIT's and MassGIS's experimental orthophoto browser for Boston and Portugal's orthophoto collection (see Section 2.7). Orthophotos possess the geometric rigour of a map, while displaying the realistic views associated with photography (Fig. 2.13).

Colour infrared aerial photos are used to analyse vegetation cover (Arnold, 1997). USGS, through its National Aerial Photography Program (NAPP), systematically collects infrared aerial photos at a 1 : 40.000 scale of the entire country every five years. A complete coverage of infrared aerial photos for the country of Portugal is also freely available at the National Geographic Information System site (Section 2.7).

2.3.2 Extraction of Digital Terrain Models

Digital elevation models (DEMs) provide a realistic representation of the terrain and are used in environmental modelling and visualisation projects as discussed in Chapters 4 and 5. Elevation may be derived from stereoscopic images or single aerial photos. Estes and Lawless provide a review of these methods in the Remote Sensing Core Curriculum (see the Web address in Section 2.7).

DEMs are usually derived from stereo pairs of aerial photographs and satellite images. The method was inspired by the recognition that human inter-pupillary distance enables depth perception. In remote sensing, this distance is

Fig. 2.13 Example of an orthophoto from Portugal's SNIG collection
Source: http://snig.cnig.pt. Image courtesy of the National Center for Geographic Information

exaggerated to become the length between the two consecutive exposures needed to obtain the stereo pairs. These should overlap by at least 50%. Figure 2.14 illustrates the concept.

The calculation of the height of a point on the earth's surface depends on the flying height, distance between exposures, focal length of the camera, and the parallax associated to each point. The focal length of the camera is the distance from the focal plane to the centre of the lens when focused at infinity. The focal length of a camera should be selected according to the type of terrain. For instance, wide-angle lenses (short focal length) are suited for photographing flat terrain as they exaggerate the displacement of tall objects.

Image parallax is the change in position of an image from one photograph to the next. Rao *et al.* (1996) recognised that there are two types of parallax associated with stereo images obtained remotely:

- Vertical parallax (*Y*-parallax), due to the vehicle motion errors.
- Horizontal parallax (*X*-parallax) due to terrain relief.

Before assessing *X*-parallax, the two images (the stereo pair) are aligned along the horizontal axis to remove the *Y*-parallax. Then, the horizontal parallax can

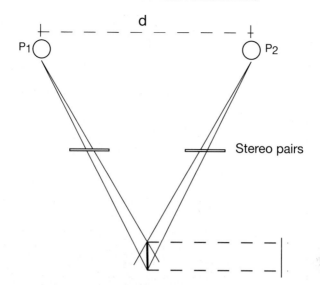

Fig. 2.14 Perception of depth in human vision derives from intra-pupillary distance. Substituting human pupils by cameras and increasing the distance between them is the starting point for obtaining height measurements from stereo pairs

be computed and, thus, digital elevations. This computation is possible because the position of the camera and the direction that the camera is pointing for each image is known at all times (Wolff and Yaeger, 1993). Knowing the elevations for each pixel, digital elevation models can be computed. Several companies specialise in extracting DEMs from remotely sensed data (IRS, SPOT, RADARSAT) and now offer their services on the Internet (see Section 2.7).

Single aerial photos can also be applied to extract digital elevation models, as the height of an object on the ground can be determined from photos taken from a low-flying aircraft. The method shown in Fig. 2.15 takes advantage of the geometric relationships between the focal length of the camera's lens, the horizontal spatial resolution of the film, the altitude of the plane of the lens, and the height of the object on the ground (Sabins, 1997).

Photos are usually available in the JPEG format (see Chapter 3) on the Web. LizardTech's MrSID is a proven alternative for large image database backed sites (see Section 2.7). MrSID is based on a wavelet compression algorithm.

2.3.3 Ground Photography

Ground photos have been used for urban planning as demonstrated by Owens (1993) and landscape analysis (Kent and Elliot, 1995). Ground photos were

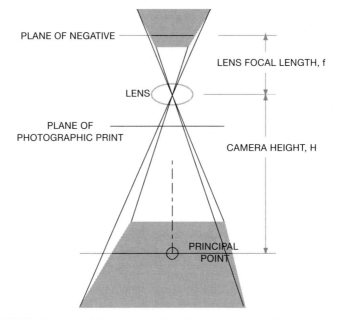

PLANE OF NEGATIVE

LENS FOCAL LENGTH, f

LENS

PLANE OF
PHOTOGRAPHIC PRINT

CAMERA HEIGHT, H

PRINCIPAL
POINT

Fig. 2.15 Derivation of the height of an image feature taken by a camera of focal length (f) and a height (H) from the ground

From Sabins (1997, fig. 2-9), with permission

utilised extensively in environmental studies for Expo98, and are described in Chapter 7. Videos included on the book's Web site show the use of photos in monitoring and impact assessment activities associated with the above studies.

Immersive photography, a development from traditional photos, is based on the creation of 360-degree panorama environments in indoor or outdoor environments. It is becoming increasingly popular on the Web. There are two main tools to develop these environments: Apple's QuickTime VR and Interactive Pictures Corporation's IPIX. QuickTime VR relies on the 'stitching' of individual images into a single file. IPIX requires a fisheye lens attached to a camera. Figure 2.16 illustrates the concepts underlying IPIX (see also a demonstration on the book's Web site). Each 360-degree panorama is essentially a node. To cover an area, several nodes may be used. Multinode photographic environments are represented in two-dimensional maps to facilitate navigation. Immersive photos enable space exploration by just dragging the mouse. They can also have hot spots leading to interactive objects.

Photography and computer graphics techniques are being applied to derive three-dimensional models of structures in the environment by the MIT City Scanning Project. This project uses a high-resolution digital camera on a mobile

Fig. 2.16 Two base hemispheric photos which will lead to an IPIX
360-degree panorama

Image courtesy of Interactive Pictures Corporation

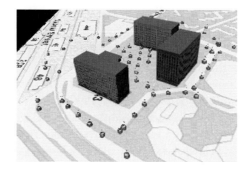

Fig. 2.17 Nodes for scanning Tech Square in Cambridge, Massachusetts

Image courtesy of Seth Teller, MIT Computer Graphics Group

platform (see the Web site address in Section 2.7). It includes instrumentation to estimate, for each image acquired, the position and orientation of the camera. A new image is recorded every few seconds. A companion vehicle is utilised to acquire geo-referenced video images of interiors and exteriors. The images and other records allow the development of a three-dimensional model of a site. Users can then view the model from any perspective and under different simulated lighting conditions. Figure 2.17 illustrates the points where images had to be taken to derive a three-dimensional model of Tech Square which is close to the MIT campus.

2.4 Videography

Video has been used in environmental monitoring both in the form of an airborne device and as a ground sensor. Video is a sequence of images, usually called frames. Important features of video are the frame rate and number of scanning lines (rows of pixels). In North America, the video standard is the NTSC format (30 frames per second, 525 lines). In most of Europe, PAL is the standard format with 25 frames per second and 625 scanning lines.

The frame rates and resolution of video in the Internet are dependent on the bandwidth available and video formats. Popular Internet video formats are QuickTime, MPEG, Real Video, and Windows Media.

Video has been used in a variety of environmental applications such as:

- The evaluation of monoxide carbon emissions from vehicles (Stephens and Caddle, 1991).
- Automobile traffic (Kilger, 1992) and pedestrian traffic analysis (Rourke and Bell, 1992), something that is now common in the Internet through the use of Web cameras.
- Coastal management (Raper and McCarthy, 1994).
- Environmental impact assessment using multispectral videography (Snider *et al.*, 1994).
- Analysis of a jet plume (Vauquelin, 1996).
- Assessment of pipeline environmental impact (Um and Wright, 1996).
- Classification of wetlands (Seibert *et al.*, 1996).
- Classification of riparian systems using multispectral data (Neale, 1997).
- Monitoring of industrial emissions using a combination of LIDAR (a laser sensor described below) and videography (Weibring *et al.*, 1998).
- A system to estimate parameters for air pollution models (Ferreira, 1998), described in Box 2.2.

Television stations commonly use video to report on problems such as:

- Water pollution (floating materials, foam, oil, colour).
- Air pollution (plumes, smog conditions).
- Solid waste (accumulating trash in urban areas, landfill conditions).
- Landscape deterioration and catastrophes such as floods, forest fires, oil spills, earthquakes, volcanic eruptions, and hurricanes.

Nature videos are also relevant documentaries on many ecosystems, animal, and plant species around the world. Discovery Channel, National Geographic, and the BBC nature programmes, among others, are illustrative examples. The movie and computer graphics industries have also shown the use of video to capture animal movement in order to improve digital animation (see Chapter 5). However, most environmental professionals and scientists have neglected the use of video images, and Chapter 4 attempts to show how video can be used to develop and visualise the results of environmental simulation models.

Airborne videography is the major professional application of video in environmental analysis. Vlcek (1998) was one of the first to propose this as an

Box 2.2 *Estimating Parameters for Air Quality Models Using Video Images*

Ferreira (1998) has developed a method using inexpensive cameras to evaluate vapour plumes from a power plant. The location of the two cameras used in the study is shown in Fig. 2.18. Figure 2.19 shows how this system can easily be used to evaluate the height of the plumes. Finally, Fig. 2.20 displays the image processing methods applied to calculate wind direction. Adobe Photoshop filters were applied in this case. Height of the plumes and wind direction are parameters used in classic air quality models.

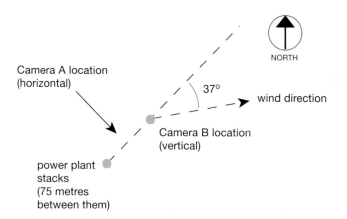

Fig. 2.18 Location of video cameras recording vapour plumes from power plant stacks

Source: Ferreira, 1998

Fig. 2.19 Calculation of the vapour plume rise

Source: Ferreira, 1998

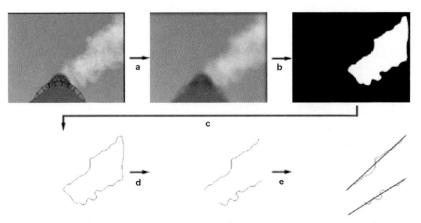

Fig. 2.20 Image processing operations for calculation of the wind direction
from the vertical camera near the stack: (a) Gaussian blur; (b) threshold;
(c) trace contour; (d) selection of the contours relative to directions;
(e) computation of corresponding vectors using linear regression

Source: Ferreira, 1998

alternative to remote sensing methods. Mausel *et al.* (1992) considered that
videography has the following advantages:

- low cost;
- real-time or close to real-time availability of imagery for digital processing;
- real-time or near real-time availability of images for visual assessment;
- ability to collect spectral data in the very narrow bands in the visible to near
  infrared and mid infrared water absorption regions;
- data redundancies as images are acquired every 30th of a second for the
  NTSC system and 24th of a second for the PAL system. This redundancy
  produces multiple views.

The most significant disadvantage of video over aerial photography, pointed out
by Mausel *et al.* (1992), is its lower resolution. However, Seibert *et al.* (1996)
reported that even using an inexpensive camera (three charge-coupled device
sensors, Hi-8 format, fast shutter speeds), the results were comparable to the
ones obtained with 35 mm slide film for the purpose of classifying a wetland.

In many regions of the globe, one can see the advantages of inexpensive
videography in surveying flights such as the ones described by Clarke (1986).
These flights termed Systematic Reconaissance Flights cover regions, divided
into 10 km² grids, use planes flying at 150 km per hour at an altitude of 100 m.
Information derived from such airborne surveys include:

- estimates of number and density of major domestic and wild herbivores;
- a profile of the seasonal movements of major herbivores;
- a vegetation map;
- a soil map expressed in terms of soil colour types;
- an outline of areas that are important to wildlife;
- an outline of livestock areas;
- a land use map.

2.5 Sound

Sound is relevant for environmental professionals and scientists for two reasons: the environment is populated by sound, both unwanted (noise) or pleasant, such as many natural sounds; the digital systems representing environmental phenomena use sound to reproduce environmental sounds and to improve their user interface.

Sound consists of pressure waves travelling through air with a frequency between 20 and 20,000 Hertz. Any sound can be characterised as the super-imposition of sine waves (Fig. 2.21) with different frequencies and amplitudes (Loucks *et al.*, 1973). Sound frequency refers to how quickly the air vibrates. With higher frequencies, sound waves are closer to each other. Frequency is felt as the 'pitch' of a sound, or its 'highness' (i.e., as a piccolo) or 'lowness' (i.e., as a tuba) (Scaletti and Craig, 1993). Sound amplitude refers to the amount of pressure exerted by the air. It can be seen as the height of the wave (Fig. 2.21). Amplitude is described in units of pressure per unit area, and it is felt as the loudness of a sound. As the pressure measures cover a broad range, they are not used. Instead, a logarithmic scale of decibels (dB) has been adopted. A sound level is given by:

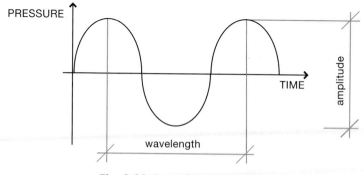

Fig. 2.21 Sound wave function

Table 2.4 Decibel levels

140	threshold of pain
130	riveting on steel plate
120	pneumatic drill
110	loud car horn at 1m
100	alarm clock at 1m
90	inside underground train
80	inside bus
70	street corner traffic
60	conversational speech
50	business office
40	living room
30	bedroom at night
20	broadcasting studio
10	normal breathing

Source: Adapted from Newson, 1992.

$$L = 10\log_{10}(P/\mathrm{p})^2 \, \mathrm{dB}$$

Where P is the amplitude of pressure fluctuations, and p is $20\,\mu\mathrm{Pa}$, which is considered to be the lowest audible sound. Table 2.4 shows decibel levels for a number of familiar environments.

The most obvious characteristic of noise (or unwanted sound) is loudness, but the annoyance it may cause is also a function of the frequency of the noise (Loucks *et al.*, 1973). Given equal noise levels, a high-frequency noise is usually considered more annoying than a low-frequency sound. Combinations of pressure and frequency are synthesised in an *A*-weighted decibel scale or dB(*A*). This scale tries to model the sensitivity of the human ear (or the average sensitivity, due to the variations among individuals). Another feature of sound is timbre, the general prevailing quality or characteristics of a sound. Scaletti and Craig (1993) illustrate the concept by pointing out that a saxophone does not sound like a violin when playing the same note. This is because the shape of the corresponding sound waves is different. From an environmental standpoint, Krygier (1994) also suggests other variables to describe sound such as:

- The location of a sound in a two or three-dimensional space.
- The length of time a sound is or is not heard.
- The relation between the duration of sound and silence over time.
- The sequence of sounds over time.
- The time it takes a sound to reach its maximum or minimum intensity level.

Noise has been extensively studied in the context of residential areas (Fields, 1998), impacts of railways (Kurze, 1996), and airport and aircraft noise (Attenborough, 1998, Zaporozhets and Tokarev, 1998), among other areas.

An ongoing unusual noise study is the Netherlands' monitoring plan for the areas surrounding Shipol Airport as part of the study for the New Airport. In this study, the public is welcome to complain about noise and can call a free telephone number. Some people are actually involved in measuring noise with devices they purchased. The consumption of aspirin in local pharmacies is also being recorded and can be interpreted as indirect public participation (Scholten, 1998). Governmental measurements and public involvement (telephone calls, measurements, and aspirin consumption) have been introduced in a geographical information system.

The collection of environmental sounds is not limited to noise measurement. There has been an extensive work in biology and ecology, recording nature sounds. A most impressive source is Cornell University's Library of Natural Sounds (see the Internet address in Section 2.7). This library includes 130,000 wildlife recordings of more than 6000 species of birds in addition to recordings of sounds of reptiles, amphibians, and mammals. The BBC Supersense Series on animal vision and hearing systems is also a good source of information (Downs, 1988). These collections may help the development of multimedia environmental impact documents. These documents can illustrate the effects of a project in the sounds of a natural environment.

Sound can also be used in environmental decision support systems to represent abstract data through sound, convey system status information, and warn the user (Buxton, 1989). Chapters 5 and 6 report illustrative applications.

2.6 On-line Sensors

Although *Star Trek*-type use of sensors may still be years away, sensors are thought to be the next wave of innovation by Saffo (1997). There is indeed a growing market for inexpensive and wire-less sensors for home use. The Nighthawk Carbon Monoxide Detector sold through utility companies in the United States is a prime example. The use of sensors for professional and scientific environmental monitoring is also expanding. By using transducers with digital communication and graphics capabilities, ground sensor data may be easily integrated in real time in environmental multimedia information systems.

Campbell (1997) edited a book reporting on the state-of-the-art in sensor systems for environmental monitoring including fibre–optic, optic, laser, electronic noses, and biosensor devices. Additional reviews may be found in Rogers (1995) on the use of biosensors; Rogers and Poziomek (1998) for fibre-optic sensors;

Bartlett *et al.* (1997) for machine olfaction; Choudhury (1998) on optical sensors; and Mason *et al.* (1998) for portable sensors. Wilson provided a novel view of sensors by proposing the use of seabirds (1992). Humans are also natural environmental sensors and their capabilities are discussed in Section 2.6.7.

2.6.1 Fibre-optic Sensors

Fibre-optic sensors enable the transmission of light over long distances (hundreds of metres). Thus, a sensor head may be located at a distance from the instrumentation, which makes them suitable for harsh environments (Stewart, 1997). The intensity, wavelength, phase, and polarisation of light can be used as measurement parameters. Several wavelengths can be launched in both directions, which allows the possibility of monitoring several chemicals and the temperature with the same sensor. As multiplexing of fibre-optic systems is feasible, expensive instrumentation can be shared among a number of sampling sites (Stewart, 1997). Fibre-optic sensors have been used to monitor (Stewart, 1997):

- *Air pollution.* Carbon monoxide has been measured by direct absorption using the infrared or the near-infrared bands.
- *Seawater.* Using absorbing and fluorescent indicator dyes, pH, and dissolved oxygen have been measured.
- *Groundwater and drinking water contamination.* Chloroform and trichloroethylene have been measured by using sensors based on pyridine which absorbs light in the green region.

2.6.2 Optical Sensors

Integrated optics is a recent technique in environmental sensing. Optical sensors rely on the use of beams of light. Thus, electrical connections are replaced by optical waveguides and semiconductor integrated circuits by optical circuits (Magill, 1997). Optical sensing relies on detecting an alteration in an optical signal as a result of a change occurring at the site being sensed. Optical parameters used include absorbency, refractive index change, or fluorescence (Magill, 1997). These sensors may be applied for gases (e.g., carbon dioxide, carbon monoxide, oxides of nitrogen and sulphur) and water (e.g., phosphates, heavy metals). Integrated optical sensors present advantages in robustness and miniaturisation, which may lead to future commercial applications.

2.6.3 Laser Sensors

Laser sensors defy the science fiction vision of light and portable sensors—they are bulky and expensive. They are based on laser mass spectrometry that can be applied to any sample, solid, liquid, or gas. They use laser beams in conjunction with mass spectrometers, nuclear detectors, and gas chromatographic techniques (Ledingham and Campbell, 1997). Typically, laser beams hit targets and the backscattered light is collected and processed.

LIDAR (LIght Detection And Ranging), the optical equivalent of RADAR (radio beam), is one of the most used laser sensors. It is used for monitoring tropospheric and stratospheric pollutants for distances up to hundreds of kilometres. LIDAR has been also used to measure heights of forest canopies and water depth relative to the water surface.

2.6.4 Electrochemical Sensors

These sensors are based in microelectrodes and can be battery operated, which makes them portable. They have been used mainly in water analysis. Electrode fouling and polishing is a problem hampering practical application of these sensors (Ansell and McNaughton, 1997).

2.6.5 Electronic Noses

Electronic noses (Bartlett *et al.*, 1997, Bates and Campbell, 1997; Pearce, 1997) rely on the use of gas sensors coupled with pattern recognition techniques to identify odours. Gas sensors are often based on piezoelectric sensors (Hepher and Reilly, 1997). Piezoelectric materials are crystals possessing a structure lacking a centre of symmetry when subject to mechanical strain.

2.6.6 Biosensors

Biosensors are devices that transform biological reactions into electrical signals (Cardosi and Haggett, 1997). They rely on sensing biological elements that may include tissues, cells, organelles, membranes, enzymes, receptors, antibodies, or nucleic acids. These elements work with transducers based on a variety of principles ranging from potentiometry to accoustics. Biosensors are specific for a particular chemical or group of related chemicals.

2.6.7 Human Sensory Information

Humans continuously monitor the environment through sensations that lead to the perception of problems (Goldstein, 1999). Vision enables the identification of water and air pollution, solid wastes, and landscape deterioration, as discussed in Section 2.4 on videography. Noise is captured by hearing. Olfaction and taste detect chemicals present in the water and air. Skin becomes irritated under adverse environmental conditions.

As environmental sensors, humans are subjective which limits their credibility: environmental conditions that are acceptable for some become totally unacceptable to others. This subjectivity is due to differences in age, gender, type of occupation, education, previous illnesses, medication, and other factors as discussed by Molhave *et al.* (1991). Empirical evidence also shows that there is no significant correlation between perceptions derived from tasting and smelling potential water contaminants and their toxicity (Young *et al.*, 1996): some chemicals were detected at concentrations far below those which are of concern; and the non-existence of organopletic effects does not guarantee the absence of contaminants above the desirable limits.

Even with these limitations, complaints by the public usually reflect poor environmental conditions. The study done by Young *et al.* (1996), shows how such complaints can be finely tuned to actually progress towards an improved identification of contaminants. Those authors listed the most frequently used odour and taste descriptors associated to several water contaminants.

Young *et al.*'s (1996) work is a methodical study on the application of two senses to the evaluation of drinking water. The research on the combined use of all our senses (a classical multisensor fusion problem) in evaluating other environmental media is a logical development.

2.7 Further Exploration

There are a large number of Web sites related to environmental data sources and acquisition methods mentioned in this chapter. The following list, including relevant sites, is periodically updated at (http://gasa.dcea.fct.unl.pt/camara/index.html). At this site, student projects are also suggested to further explore these topics.

General Environmental Information Resources

For a general collection on environmental information resources, investigate the institutional links referred to in Chapter 1 and:

- National Environmental Data Index, a gateway to environmental information produced by Departments and Agencies of the US Government (http://www.nedi.gov).
- This site uses information locator services. An example is the US Environmental Protection Agency (EPA) locator service (http://www.epa.gov/gils).
- Centre for International Earth Science Information Network (CIESIN) that provides access to metadata, data, software, and services for earth scientists (http://www.ciesin.org).
- University of California at Davis maintains the Information Center for the Environment (http://ice.ucdavis.edu).

- A search engine for environmental information is provided by the Geoindex service (http://www.geoindex.com).

Natural and Man-made Disasters

Sites related to natural and man-made disasters include:

- United States Geological Survey (USGS) site is a reference on earthquakes, landslides, floods, and volcanoes occurring around the world (http://www.usgs.gov).
- Charles Surt University in Australia hosts a virtual library on fires (http://www.csu.edu.au/firenet).
- Tromso Satellite Station, Norway, provides services on oil spill detection using radar images from satellite and aircraft (http://www.tss.no/services/index.html).

Media-specific Environmental Problems

Water quality related data and software may be accessed at the EPA and USGS sites (http://www.epa.gov and http://www.usgs.gov). The Water Environment Federation site (http://www.wef.org) has extensive links to sites on biosolids, watershed, and industry issues. In addition, it has a comprehensive list of Web addresses for governmental agencies throughout the world.

The EPA site (http://www.epa.gov/epahome/programs.htm) covers air quality, soil contamination, and hazardous waste related problems.

Soil erosion related data and methods may be found at the US National Soil Erosion Research Laboratory site (http://topsoil.nserl.purdue.edu/default.htm).

The Ecology Society of America maintains an excellent site with links to electronic publications, data, and organisational sites related to ecology (http://esa.sdsc.edu).

Ecology Communications maintains a site dedicated to the general public (http://www.ecology.com).

United States and international climate data may be accessed at the National Weather Service (http://www.nws.noaa.gov).

Collaborative Monitoring

Environmental monitoring may be a collaborative effort as shown by EPA's citizen participation programmes (http://www.epa.gov/epahome/citizen.htm). The Globe initiative including students from around the world is another illustrative example (http://www.globe.gov).

General Sites on Remote Sensing

The Centre for Earth Observation (CEO) provides extensive information on remote sensing activities (http://infeo.ceo.org). This site includes case studies on the application of remote sensing to change detection (protected areas, desertification), land cover, erosion, landslides, flooding, vulcanology, forest fire, and renewable energy issues.

Four other comprehensive remote sensing sites are offered by Canadian Center for Remote Sensing (http://www.ccrs.nrcan.gc.ca), the NASA Destination Earth (http://www.earth.nasa.gov), USGS Earth Resources Observation System (http://edchttp://www.cr.usgs.gov), and Imagenet (http://www.imagenet.com). The Canadian site includes a tutorial on remote sensing and an image showcase. Destination Earth provides links to many of the major earth observation projects. The USGS EROS system provides access to satellite images and aerial photography. Imagenet is a service for a consortia of remote sensing companies.

An archive of remote sensing images may be found at the Joint Propulsion Laboratory site (http://www.jpl.nasa.gov/pictures) and the NASA Observatorium (http://observe.ivv.nasa.gov/nasa/education/tools/stepby/multi.html). The latter provides free LANDSAT images for educational purposes.

Other archives with images on environmental problems are located at the USGS Earthshots site (http://www.usgs.gov/earthshots) and National Oceanic and Atmospheric Administration's National Environmental Satellite Data and Information Service (http://www.nesdis.nooa.gov).

Specific Remote Sensors

Information on specific remote sensors can be found at the following sites:

- LANDSAT programs are at (http://LANDSAT.gsfc.nasa.gov). The SPOT program site is at (http://www.spot.com).
- Joint Propulsion Lab hosts the Airborne Visible Infrared Imaging Spectrometer (AVIRIS) program (http://makalu.jpl.nasa.gov).
- The Geostationary Operational Environmental Satellite (GOES) related observation programs can be followed at (http://goes1.gsfc.nasa.gov).
- Sensors directed towards oceanic problems include SeaWiFS (http://seawifs.gsfc.nasa.gov/SEAWIFS.html) and TOPEX-POSEIDON (http://topex-www.jpl.nasa.gov/). The latter is non-imaging radar system.
- High-resolution spatial sensors are offered by Earth Watch (http://www.digitalglobe.com), Space Imaging (http://www.spaceimaging.com), SPIN-2 (http://www.spin-2.com), and Orbimage (http://www.orbimage.com).
- Canada's Radarsat program is shown at (http://radarsat.space.gc.ca). This site includes the usual gallery of images and video, complemented by games. The European Space Agency's ERS-1 and ENVISAT programs are located at (http://www.esa.int/esa/progs/eo.html).
- Other specific sensors are described in review documents such as those found at

(http://www.gaf.de/Train4DM/contents/access_eo_information.pdf) and (http://my.netian.com/~kiddew/spacerace.htm).

Image Processing

Adobe PhotoShop (http://www.adobe.com) is widely used for image processing. However, for processing spatial images there are specialised software applications (Jensen, 1996) that perform tasks such as:

- Pre-processing, including radiometric and geometric correction.
- Display and enhancement functions such as image algebra, spatial filtering, principal component analysis and Fourier transforms.
- Information extraction such as supervised and unsupervised classification and extraction of digital terrain models and orthophotos.
- Integration with geographical information systems.

Vendors include:

- ERDAS (http://www.erdas.com).
- ER Mapper (http://www.ermapper.com).
- Genasys (http://www.genasys.com).
- Intergraph (http://www.intergraph.com).
- Microimages (http://www.microimages.com).

Many geographical information systems also include image-processing capabilities such as IDRISI (http://www.clarklabs.org) and GRASS (http://www.geog.uni-hannover.de/grass). Researchers from Purdue University have developed MultiSpec for processing multispectral and hyperspectral images at (http://dynamo.ecn.purdue.edu/%7Ebiehl/MultiSpec/Index.html).

Most of these vendors as well as remote-sensing firms, such as SPIN-2, also enable the generation of Digital Elevation Models (DEMs) from remote sensed data.

Geo@News is a project that shows the application of simple image-processing techniques to enhance satellite and aerial photograph images to be used by the media (http://gasa.dcea.fct.unl.pt/geo@news/index.html).

Global Positioning System Data

A generic site on GPS data and applications is provided at (http://www.gypsy.com/gpsinfo). A more research-oriented GPS site is offered at Iowa State University (http://www.gpsworld.com).

Two non-traditional GPS applications include water vapour measurements (http://www4.etl.noaa.gov/gps) and an analysis of the earth's tectonic plates (http://sideshow.jpl.nasa.gov/mbh/series.html).

Photography

Philip Greenspun's Photo.Net (http://photo.net) provides a wealth of material on the use of photos on the Net. Relevant sites on the use of photography for environmental applications include:

- The MIT orthophotos browser (http://ortho.mit.edu/nsdi). The National System for Geographic Information of Portugal also offers orthophotos on the Web (http://snig.cnig.pt). This collection uses the MrSID format based on the use of the Discrete Wavelet Transform (see Appendix 2), proposed by LizardTech (http://www.lizardtech.com). For a comparison of MrSid with JPEG and FlashPix formats (http://www.digitalimaging.org), see a technical paper at LizardTech's site.
- For a list of applications and resources on aerial photography (http://mollisol.agry.purdue.edu/~helt/aerial.html). For resources on unmanned aerial photography, a useful site may be found at (http://http://www.auvsi.org/auvsicc/).
- Immersive photography sites include QuickTime VR (http://www.apple.com/quicktime/authoring/vrfaq.html) and IPIX (http://www.ipix.com). The Virtual Parks site (http://www.virtualparks.org) is an application of immersive photography to natural areas.
- The MIT City Scanning Project description and images is available at (http://graphics.lcs.mit.edu/city/city.html).

Videography

The generic sites on the use of video in the Internet are QuickTime's (see http://www.apple.com), Real Video (http://www.realvideo.com), Windows Media (http://www.microsoft.com), and MPEG (http://www.cselt.it/mpeg). The location of Web cameras around the world is shown at (http://www.camcentral.com/location.html).

Sound

The Institute for Noise Control Engineering (http://www.ince.org) is a reference on noise pollution and control. For noise cancellation see (http://www.nct-active.com). A public interest page on noise is maintained at (http://www.nonoise.org).

For Cornell's Library of Natural Sounds see (http://birds.cornell.edu/LNS/).

Sensors

Information on environmental sensors may be found from these vendors:

- Environmental Sensors at (http://www.envsens.com).
- Imagiworks at (http://www.iamgiworks.com), which is specialized in connecting sensors to mobile devices such as Palm Pilots.

• CyranoSciences at http://www.cyranosciences, specialised in portable electronic noses.

A relevant source is NASA's site at (http://csmt.jpl.nasa.gov/environmental.html).
Information on the application of LIDAR sensors for mapping may be found at (http://www.airbornelasermapping.com).

References

ACS (American Chemical Society) (1998). *Principles of Environmental Sampling*. New York: ACS.

American Society of Photogrammetry and Remote Sensing (ASPRS) (1996). *Earth Observing Platforms and Sensors*, CD-ROM, Version 1.0. New York: ASPRS.

Ansell, R.O. and McNaughton (1997). Electrochemical Sensors. In M. Campbell (ed.), *Sensor Systems for Environmental Monitoring*, 100–26. London: Chapman & Hall.

Arnold, R.H. (1997). *Interpretation of Airphotos and Remotely Sensed Imagery*. Upper Saddle River, NJ: Prentice Hall.

Attenborough, K. (1998). Special Issue: Airport and Aircraft Noise Modelling and Control. *Applied Acoustics*, 55/2: 87–7.

Bartlett, P.N., Elliot, J.M., and Gardner, J.W. (1997). Applications of, and Developments in, Machine Olfaction. *Annali di Chimica Roma*, 87/1–2: 33–44.

Bates, J.R. and Campbell, M. (1997). Gas sensors and analysers. In M. Campbell (ed.), *Sensor Systems for Environmental Monitoring*, 127–78. London: Chapman & Hall.

Buxton, W. (1989). Introduction to this special issue on nonspeech audio. Human Computer Interaction, 4/1: 1–9.

Campbell, M. (ed.) (1997). *Sensor Systems for Environmental Monitoring*. London: Chapman & Hall.

Canadian Center for Remote Sensing (1998). Tutorial on Remote Sensing. Available at (http://www.ccrs.nrcan.gc.ca).

Cardosi, M. and Haggett, B. (1997). Biosensor Devices. In M. Campbell (ed.), *Sensor Systems for Environmental Monitoring*, 210–67. London: Chapman & Hall.

Carlson, G. and Patel, B. (1997). A New Era Dawns for Geospatial Imagery. *GIS World*, 10/3: 36–40.

Choudhury, P.K. (1998). Optical Sensors for Environmental Monitoring. *Current Science India*, 74/9: 723–5.

Clarke, R. (ed.) (1986). *The Handbook of Ecological Monitoring*. Oxford: Clarendon Press.

Cohen, D.J., Jensen, J.R., and Bresnahan, D.J. (1995). The Design and Implementation of an Integrated Geographic Information System for Environmental Applications. *Photogrammetric Engineering & Remote Sensing*, 61/11: 1393–404.

Cox, L.H. and Piegorsch, W.W. (1996). Combining Environmental Information: I. Environmental Monitoring, Measurement and Assessment. *Environmetrics*, 7: 299–308.

Downs, E. (1988). *Supersense*. London: BBC (video).

Ferreira, F. (1998). Digital Video Applied to Air Pollution Emission's Monitoring and Modelling. PhD Dissertation, New University of Lisbon, Monte de Caparica, Portugal.

Fields, J.M. (1998). Reactions to Environmental Noise in Residential Areas. *Journal of the Accoustical Society of America*, 104/4: 2245–60.

Foody, G.M. and Curran, P.J. (eds) (1994). *Environmental Remote Sensing: From Regional to Global Scales*. Chichester: Wiley.

Frost, A.R., Schofield, C.P., Beaulah, S.A., Mottram, T.T., Lines, J.A., and Wathes, C.M. (1997). A Review of Livestock Monitoring and the Need for Integrated Systems. *Computers and Electronics in Agriculture*, 17/2: 139–59.

Gibbons, G. (1992). The Global Positioning System as a Complementary Tool for Remote Sensing and Other Applications. *Photogrammetric Engineering and Remote Sensing*, 58/8: 1255–7.

Gilbert, R.O. (1987). *Statistical Methods for Environmental Pollution Monitoring*. New York: Van Nostrand Reinhold.

Goldstein, E.B. (1999). *Sensation and Perception*. Pacific Grove, CA: Brooks/Cole.

Gorman, L., Morang, A., and Larson, R. (1998). Monitoring the Coastal Environment, Part IV: Mapping, Shoreline Changes and Bathymetric Analysis. *Journal of Coastal Research*, 14/1: 61–92.

Gower, J.F. and Borstad, G.A. (1990). Mapping of Phytoplankton by Solar-Stimulated Fluorescence Using an Imaging Spectrometer. *International Journal of Remote Sensing*, 11/2: 313–20.

Harmancioglou, N.B., Singh, V.P., and Alpaslan, M.N. (eds) (1998). *Environmental Data Management*. Dordrecht: Kluwer.

Hepher, M.J. and Reilly, D. (1997). Piezoelectric Sensors. In M. Campbell (ed.), *Sensor Systems for Environmental Monitoring*, 179–209. London: Chapman & Hall.

Hurn, J. (1993). *Differential GPS Explained*. Sunnyvale, CA: Trimble Navigation.

Jensen, J.R. (1996). *Introductory Digital Image Processing*. Upper Saddle River, NJ: Prentice Hall.

Jones, L. (1992). *TOPEX/POSEIDON—Oceanography from Space: The Oceans and Climate*. Washington, DC: National Aeronautics and Space Administration.

Kent, R.L. and Elliot, C.L. (1995). Scenic Routes Linking and Protecting Natural and Cultural Landscape Features: A Greenway Skeleton. *Landscape and Urban Planning*, 33/1–3: 341–55.

Kilger, M. (1992). Video-Based Traffic Monitoring. *Proceedings of the IEEE 4th International Conference on Image Processing and Its Applications*, New York, 83–92.

Knott, J.M., Wenner, E.L., and Wedt, P.H. (1997). Effects of Pipeline Construction on the Vegetation and Macrofauna of two South Carolina, USA, Salt Marshes. *Wetlands*, 17/1: 65–81.

Krygier, J.B. (1994). Sound and Geographic Visualization. In A. MacEachren and D.R.F. Taylor (eds), *Visualization in Modern Cartography*. Oxford: Pergamon Press.

Kurze, U.J. (1996). Tools for Measuring, Predicting and Reducing the Environmental Impact from Railway Noise and Vibration. *Journal of Sound Vibration*, 193/1: 237–51.

Lancia (1993). TV commercial for Lancia, presented at Imagina 93, Monte Carlo (video).

Ledingham, K.W.D. and Campbell, M. (1997). Laser Sensors. In M. Campbell (ed.), *Sensor Systems for Environmental Monitoring*, 65–99. London: Chapman & Hall.

Loftis, J.C., McBride, G.B., and Ellis, J.C. (1991). Considerations of Scale in Water Quality Monitoring and Data Analysis. *Water Resources Bulletin*, 27/1: 255–64.

Loucks, D.P., Bower, B.T., and Spofford, W.O. (1973). Environmental Noise Management. *Journal of the Environmental Engineering Division, ASCE*, 99/6: 813–29.

Magill, J.V. (1997). Integrated Optic Sensors. In M. Campbell (ed.), *Sensor Systems for Environmental Monitoring*, 41–64. London: Chapman & Hall.

Maher, W.A., Cullen, P.W., and Norris, R.H. (1994). Framework for Designing Sampling Programs. *Environmental Monitoring and Assessment*, 30: 139–62.

Mason, A., Yazdi, N., and Chavan, A.V. (1998). A Generic Multielement Microsystem for Portable Wireless Applications. *Proceedings of IEEE*, 86/8: 1733–46.

Mausel, P.W., Everitt, J.H., Escobar, D.E., and King, D.J. (1992). Airborne Videography: Current Status and Future Perspectives. *Photogrammetric Engineering & Remote Sensing*, 58/8: 1189–95.

Metcalf, J.D. and Arnold, G.P. (1997). Tracking Fish With Electronic Tags. *Nature*, 387/6634: 665–6.

Molenaar, M. (1998). To See or not To See. International Institute for Aerospace Survey and Earth Sciences (ITC), A transfer address. Enschede, The Netherlands.

Molhave, L., Jensen, J.G., and Larsen, S. (1991). Subjective Reactions to Volatile Organic Compounds as Air Pollutants. *Atmospheric Environment*, 23/7: 1283–93.

Moss, M.E. (1989). Water Quality Data in the Information Age. In R.C. Ward, J.C. Loftis, and G.B. McBride (eds), *Proceedings, International Symposium on the Design of Water Quality Information Systems*. Fort Collins, CSU Information Series, No. 61, 73–86.

Neale, C.M.U. (1997). Classification and Mapping of Riparian Systems using Airborne Multi-spectral Videography. *Restoration Ecology*, 5/4: 103–12.

Newson, M. (ed.) (1992). *Managing the Human Impact on the Natural Environment*. London: Belhaven Press.

Owens, P.M. (1993). Neighborhood Form and Pedestrian Life—Taking a Closer Look. *Landscape and Urban Planning*, 26/1–4: 115–35.

Parkinson, C.L. (1997). *Earth from Above, Using Colour-Coded Satellite Images to Examine the Global Environment*. Sausalito, CA: University Science Books.

Pearce, T.C. (1997). Computational Parallels Between the Biological Olfactory Pathway and its Analogue 'The Electronic Nose'. 2. *Biosystems*, 41/2: 69–90.

Rao, T., Rao, K.V., Kumar, A.R., Rao, D.P., and Deekshatula, B.L. (1996). Digital Terrain Model (DTM) from Indian Remote Sensing (IRS) Satellite Data from the Overlap Area of Two Adjacent Paths Using Digital Photogrammetric Techniques. *Photogrammetric Engineering and Remote Sensing*, 62/6: 727–31.

Raper, J.F. and McCarthy, T. (1994). Using Airborne Videography to Assess Coastal Evolution and Hazards. *Proceedings of EGIS 94 Conference*, Paris, 1224–8.

Rodgers, A.R., Rempel, R.S., and Abraham, K.F. (1996). A GPS-Based Telemetry System. *Wildlife Society Bulletin*, 24/3: 559–66.

Roesler, C.S. and Perry, M.J. (1995). In situ Phytoplankton Absorption, Fluorescence Emission, and Particulate Backscattering Spectra Determined from Reflectance. *Journal of Geophysical Research*, 110/C7/13: 279–94.

Rogers, K.R. (1995). Biosensors for Environmental Applications. *Biosensors Bioelectronics*, 10/6–7: 533–41.

Rogers, K.R. and Poziomek, E.J. (1998). Fiber-Optic Sensors for Environmental Monitoring. *Chemosphere*, 33/6: 1151–74.

Rourke, A. and Bell, M.G.H. (1992). Wide Area Pedestrian Monitoring Using Video Image Processing. *Proceedings of the IEEE 4th International Conference on Image Processing and Its Applications*, 563–6. New York: IEEE Press.

Rutter, S.M., Beresford, N.A., and Roberts, G. (1997). Effects of Differential Correction on Accuracy of a GPS Animal Location System. *Computers and Electronics in Agriculture*, 17/2: 177–88.

Sabins, F.F. (1997). *Remote Sensing, Principles and Interpretation*. New York: W.H. Freeman.

Saffo, P. (1997). Sensors: The Next Wave of Innovation. *Communications of the ACM*, 40/2: 93–7.

Scaletti, C. and Craig, A.D. (1993). *Using Sound to Extract Meaning from Complex Data*. Technical Report, Urbana-Champaign: CERL, University of Illinois.

Scholten, H. (1998). Noise Measurement around Shipol Airport. Personal Communication, Amsterdam.

Schott, J.R. (1997). *Remote Sensing. The Image Chain Approach*. New York: Oxford University Press.

Seibert, T.F., Sidle, J.G., and Savidge, J.A. (1996). Inexpensive Aerial Videography: Acquisition, Analysis and Reproduction. *Wetlands*, 16/2: 245–50.

Seixas, M.J. (1998). Patterns of Heterogeneity Derived from Remote Sensing Images— Implications for the Environmental Assessment of Desertification at Southern Portugal. PhD Dissertation in Environmental Engineering, New University of Lisbon, Monte de Caparica, Portugal.

Snider, M.A., Hayse, J.W., Hlohowski, I., and LaGory, K. (1994). Multispectral Airborne Videography Evaluated Environmental Impact. *GIS World*, 7/6: 50–2.

Stephens, R.D. and Cadle, S.H. (1991). Remote Sensing Measurements of Carbon Monoxide Emissions from On-road Vehicles. *Journal of Air and Waste Management*, 41/1: 39–46.

Stewart, G. (1997). Fibre Optic Sensors. In M. Campbell (ed.), *Sensor Systems for Environmental Monitoring*, 1–40. London: Chapman & Hall.

Summers, R.W. and Feare, C.J. (1995). Roost Departure by European Starlings *Sturnus vulgaris*: Effects of Competition and Choice of Feeding Site. *Journal of Avian Technology*, 26/4: 289–95.

Trujillo-Ventura, A. and Ellis, J.H. (1991). Multiobjective Air Pollution Monitoring Network Design. *Atmospheric Environment*, 25A: 469–79.

Um, J.S. and Wright, R. (1996). Pipeline Construction and Reinstatement Monitoring: Current Practice, Limitations and the Value of Airborne Videography. *Sciences of the Total Environment*, 16/3: 221–30.

Varshney, P.K. (1997). Multisensor Data Fusion. *Electronics and Communication Engineering*, 9/6: 245–53.

Vauquelin, O. (1996). Absolute Concentration Measurements Inside a Jet Plume Using Video Digitalization. *Atmospheric Environment*, 30/9: 1523–8.

Vincent, R.K. (1997). *Fundamentals of Geologic and Environmental Remote Sensing*. Upper Saddle River, NJ: Prentice Hall.

Vlcek, J. (1988). Nature of Video Images. *Proceedings of the First Workshop on Videography, American Society of Photogrammetry and Remote Sensing.* American Society of Photogrammetry and Remote Sensing (ASPRS): Falls Church, VA.

Ward, R.C. and Loftis, J.C. (1986). Establishing Design Criteria for Water Quality Monitoring Systems: Review and Synthesis. *Water Resources Bulletin, AWRA*, 22/5: 759–67.

Weibring, P., Anderson, M., and Edner, H. (1998). Remote Monitoring of Industrial Emissions by Combination of Lidar and Plume Velocity Measurement. *Applied Physics B–Lasers*, 66/3: 383–8.

Whitfield, P.H. (1988). Goals and Data Collection Design for Water Quality Monitoring. *Water Resources Bulletin, AWRA*, 24: 775–80.

Wilson, R.P. (1992). Environmental Monitoring with Seabirds—Do we Need Additional Technology. *South African Journal of Marine Science*, 12: 919–26.

Wolff, R.S. and Yaeger, L. (1993). *Visualization of Natural Phenomena.* Santa Clara, CA: Telos/Springer Verlag.

Young, W.F., Horth, H., Crane, R., Ogden, T., and Arnott, M. (1996). Taste and Odour Threshold Concentrations of Potential Potable Water Contaminants. *Water Research*, 30/2: 331–40.

Zaporozhets, O. and Tokarev, V. (1998). Aircraft Noise Modeling for Environmental Assessment around Airports. *Applied Acoustics*, 55/2: 99–127.

3

Multidimensional Environmental Information Systems

3.1 Introduction

As shown in Chapter 2, environmental data have distinct features. Gunther (1998) discussed some of these features in the context of environmental data management:

- Environmental data have large storage requirements. Just the environmentally related satellite data produced worldwide alone is at the level of a terabyte a day.
- Data are widely distributed. There are many institutions producing environmental data.
- Hardware and software supporting environmental data management is very heterogeneous (a result of the heterogeneity of institutions involved).
- Environmental data may have complex structures such as multimedia data.
- Environmental data objects often have four-dimensions (three space dimensions plus time).
- Environmental data are inaccurate due to factors such as measurement errors and the random nature of natural phenomena.

This chapter is about the environmental infrastructures that are being developed around the world to handle these issues. It will also discuss the databases, geographical, and multimedia information systems that are at the core of those infrastructures.

3.2 Databases

A data model is the structure in which a computer stores persistent information (Greenspun, 1999). In relational database management systems (RDBMS), the most widely used database systems, data models are developed from tables. An extension of the relational model called 'object-relational' for its ability to handle objects, such as multimedia data, is gaining acceptance, whereas the object-oriented model has still a relatively small market share.

3.2.1 Relational Databases

RDBMS are based on the mathematical concept of relation. A relation is defined as follows: Given sets A1, A2, . . . , An (not necessarily distinct), R is a relation on these n sets if it is a set of n-tuples, the first component of which is drawn from A1, the second component from A2, and so on. More concisely, R is a subset of the Cartesian product $A1 \times A2 \times . . . \times An$.

Frequently, relations are referred to as tables. The reason is that a relation can be readily visualised as a table with rows and columns. Each row is usually called a record (a n-tuple in mathematical terminology). Note, however, that 'relation' is not synonymous with 'table'. The difference is that a table can have duplicate rows, but a relation cannot. Throughout the text, the term 'table' is used because it gives the reader a more intuitive notion. However, when discussing a 'table', a 'relation' is meant.

The Environmental Information System developed for Expo98 is an illustrative example of a RDBMS (see Fig. 3.1). The tables of this information system

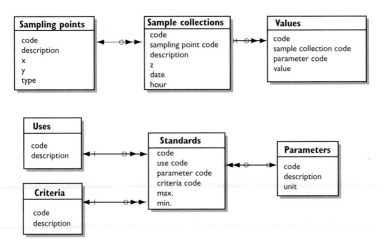

Fig. 3.1 Structure of the Expo98 Environmental Information System

are 'sampling points', 'sampling stations', 'values', 'parameters', 'standards', 'uses', and 'criteria', shown as boxes in Fig. 3.1. The entities represent the sets (or columns) of each table. The relation (or table) 'sampling points' may be then represented by the scheme (code, description, coordinate x, coordinate y, and type). Parameters reflected chemical, biological, and physical variables describing the water, air, and soil environments, energy consumption, fauna, and flora associated with Expo98's intervention site.

Note that there is always a common entity between tables, a key feature of RDBMS that enable certain operations and accessing:

- 'sampling points' and 'sample collections' is entity sampling point code;
- 'sample collections' and 'values' is sample collection code;
- 'values', 'parameters', and 'standards' is parameter code;
- 'standards' and 'uses' is use code;
- 'standards' and 'criteria' is criteria code.

There are seven basic operations with data in a RDBMS: selection, projection, Cartesian product, union, difference, join, and intersection. These operations are based on the theory of relational algebra and are nothing more than set-manipulation functions. The selection operator can be applied to find all rows in a given table that satisfy a given condition (i.e., select all sample collections at a given date). The projection operation basically allows the definition of another table that includes columns that one is interested in. The Cartesian product is the concatenation of two tables (i.e., in the Expo98 example 'sampling points' and 'sampling stations'). The union operator enables the definition of a table with rows from two tables. Two tables are difference compatible if they have the same number of columns. The difference between the two tables are the rows that are in one of the tables and not in the other. The join operator is used to integrate information from different tables. Finally, the intersection operator gives us a table with rows common to both original tables.

Relational databases are queried using a Data Manipulation Language (DML). The most popular of such languages is the Structured Query Language (SQL). A basic SQL command is of the form:

SELECT attribute 1, attribute 2 ... attribute n

FROM $R_1, R_2 ... R_K$

WHERE F

$R_1 ... R_k$ are tables

F is any formula

User interface developments further facilitate the use of SQL, as shown in examples from Expo98's Environmental Information System in Fig. 3.2. The user interface was programmed with Visual Basic 4.0 and the database engine was Access97. Microsoft, Oracle, IBM, Informix, and Sybase (see Section 3.6) develop the most popular RDBMS available on the market.

3.2.2 Object-oriented Databases

Relational databases do not handle temporary variations, such as addition or deletion of columns in relations, with ease. Relations between the contents of two or more (or parts of) tables also have to be explicitly encoded in the form of integrity constraints. RDBMS are also based on data with 'flat' structures and do not handle well non-traditional data such as multimedia data.

Instead of modelling the world as relations, it is seen from an object-oriented perspective. Then, the world becomes a collection of objects that belong to categories and include methods which are algorithms specifying how the objects in a given category are to be manipulated. Objects interact with other objects by passing messages. To structure an information system based on this

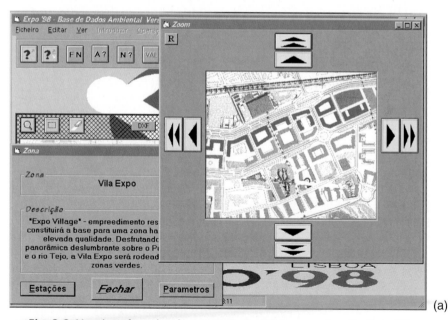

(a)

Fig. 3.2 User interface developments to facilitate accessing the Expo98 Environmental Information System: (a) navigating in space; (b) querying for parameter values; (c) values for a sampling station

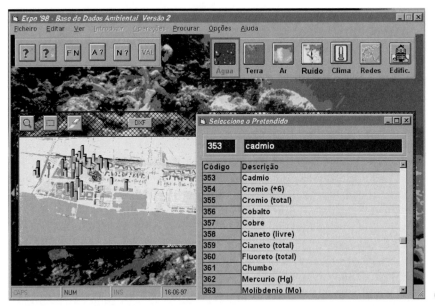

(b)

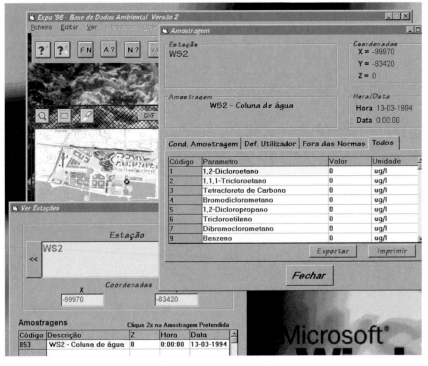

(c)

Fig. 3.2 *Continued*

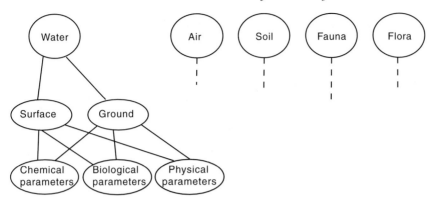

Fig. 3.3 Partial view of an object-oriented model for the Expo98
Environmental Information System

orientation, flow charts can be used showing hierarchies, as shown in Fig. 3.3.
This flow chart displays an alternative plan for the Expo98 Environmental Information System.

Most users are more comfortable with the relational model because they
understand tables better than objects. They are also interested in working
mainly at the attribute level. RDBMS are faster than object-oriented databases
in handling attributes (Greenspun, 1999). To standardise the development
and access of object-oriented systems, a group called Object Database Management Group (ODMG) has proposed an object definition language (ODL) and an
object query language (OQL) (Subramanian, 1998).

3.2.3 Object-relational Systems

Object-relational systems extend the relational model to handle complex data,
such as images. These systems are based on tables where there are object fields,
such as images, and where additional operators, such as 'match', are included.

In the Expo98 Environmental Information System, an additional object could
be the set of images associated with all the sample collections at a critical sampling point of the Tejo Estuary. By using the image of a sample, where poor
water quality would be visible, and the operator match, similar images (and associated water quality parameter values) could be retrieved. Then, one could try to
establish relationships between aesthetic considerations and water quality levels.
Leading vendors including this model are:

- Informix which has functions such as face recognition, audio data handling,
  spatial data, and video processing.
- IBM's DB2 with similar operators.

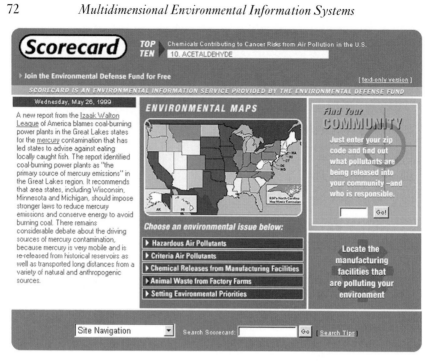

Fig. 3.4 The Scorecard Web site, May 1999

Reprinted with the permission from scorecard.org, a project of the Environmental
Defense Fund

- Oracle, which emphasises video support, text summarising, spatial data, and on-line analytical processing.

3.2.4 Web Sites Backed by Databases

Environmental databases on World Wide Web sites may provide responses to user queries. One of the most striking examples is the Scorecard database (Fig. 3.4). Scorecard using EPA's Toxic Release Inventory enables any US citizen to locate what is being released in his/her zip code area. It also shows how a chemical is used, by which industries, and how hazardous it is (Greenspun, 1999).

How does one currently access the underlying Scorecard database or any database backed Web sites? One types into a form on a Web client (i.e., the browser) that then transmits it to the Web server. This server already has an established connection to a RDBMS server and works as a RDBMS client. The RDBMS server searches the database and returns the matching data. The data then goes back from the RDBMS server to the Web server, and from there to the Web client (Greenspun, 1999). Figure 3.5 illustrates the process. Scorecard

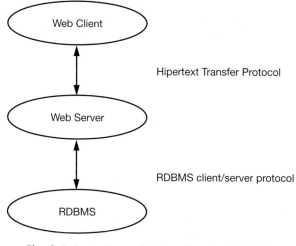

Fig. 3.5 Accessing a database backed Web site
Adapted from Greenspun, 1999

relies on the use of an Oracle database and AOL server scripts written in the Tcl language.

A working alternative to support database backed Web sites, in the Windows NT environment, is the Internet Information Server (IIS)/Active Server Page (ASP). An ASP is an HTML document containing a script that can be written using ASP code, Visual Basic, JavaScript, or other languages such as Perl. The script runs on the server, and the client receives output in the form of dynamic HTML. With ASP, you can access almost any RDBMS available in the market. Section 3.6 includes sites that provide extensive information on Active Server Page developments. Box 3.1 shows an illustrative example of an ASP application developed for the Working Site of the European Spatial Metadata Infrastructure (ESMI) project.

Box 3.1 *ASP Application to the ESMI Project*

The goal of the Web site was to facilitate the management of the ESMI project. A database with the project activities can be queried (and updated on-line) to include status reports, deadlines, and other relevant information. The structure of the database, which holds the working site information, taken from the European Spatial, Metadata Infrastructure (ESMI) working site, is shown in Fig. 3.6.

A sample of the data in the task table is shown in Fig. 3.7.

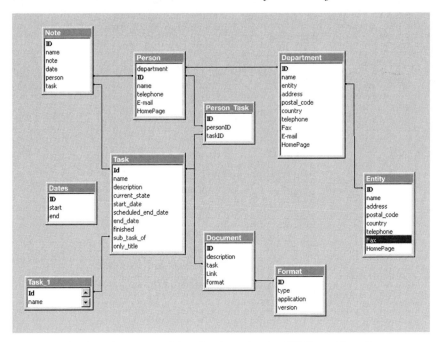

Fig. 3.6 Structure of the database for the ESMI working site

This data can be accessed with simple SQL commands with the use of Open Database Connectivity (OBDC). By using this connection, dynamic pages can be created. Part of the ASP code that generates a page showing all the current main tasks is shown in Fig. 3.8.

The complete code is processed by the HTTP server which responds to a client by giving it only HTML and JavaScript that can be read by any browser and displayed as shown in Fig. 3.9.

An example of a form to insert new data, in this case a new subtask assigned to the currently selected task 'Working Site', is provided in Fig. 3.10.

By submitting this form to the code shown in Fig. 3.11, the data is inserted into the database in real time.

The user is immediately taken to the updated page (Fig. 3.12). Note the existence of the newly created subtask 'Working Site User Tests'.

Likewise, all data in the database can be viewed and altered simply by using HTML forms that can be accessed by any member of the project from anywhere around the world, maintaining a record of all the work carried out.

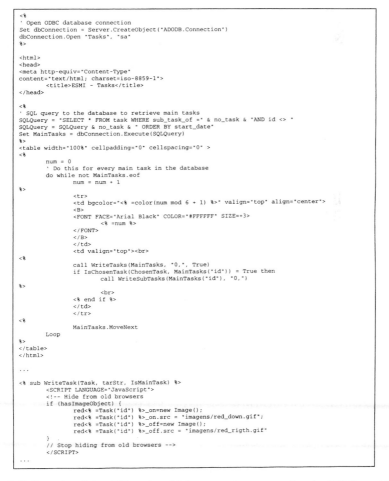

Fig. 3.7 Table of the ESMI working site database

```
<%
' Open ODBC database connection
Set dbConnection = Server.CreateObject("ADODB.Connection")
dbConnection.Open "Tasks", "sa"
%>

<html>
<head>
<meta http-equiv="Content-Type"
content="text/html; charset=iso-8859-1">
        <title>ESMI - Tasks</title>
</head>

<%
' SQL query to the database to retrieve main tasks
SQLQuery = "SELECT * FROM task WHERE sub_task_of =" & no_task & "AND id <> "
SQLQuery = SQLQuery & no_task & " ORDER BY start_date"
Set MainTasks = dbConnection.Execute(SQLQuery)
%>
<table width="100%" cellpadding="0" cellspacing="0" >
<%
        num = 0
        ' Do this for every main task in the database
        do while not MainTasks.eof
                num = num + 1
%>
                <tr>
                <td bgcolor="<% =color(num mod 6 + 1) %>" valign="top" align="center">
                <B>
                <FONT FACE="Arial Black" COLOR="#FFFFFF" SIZE=+3>
                        <% =num %>
                </FONT>
                </B>
                </td>
                <td valign="top"><br>
<%
                call WriteTasks(MainTasks, "0,", True)
                if IsChosenTask(ChosenTask, MainTasks("id")) = True then
                        call WriteSubTasks(MainTasks("id"), "0,")
%>
                <br>
                <% end if %>
                </td>
                </tr>
<%
                MainTasks.MoveNext
        Loop
%>
</table>
</html>

...

<% sub WriteTask(Task, tarStr, IsMainTask) %>
        <SCRIPT LANGUAGE="JavaScript">
        <!-- Hide from old browsers
        if (hasImageObject) {
                red<% =Task("id") %>_on=new Image();
                red<% =Task("id") %>_on.src = "imagens/red_down.gif";
                red<% =Task("id") %>_off=new Image();
                red<% =Task("id") %>_off.src = "imagens/red_rigth.gif"
        }
        // Stop hiding from old browsers -->
        </SCRIPT>

...
```

Fig. 3.8 Sample of the ASP code which generates a page for the ESMI working
site

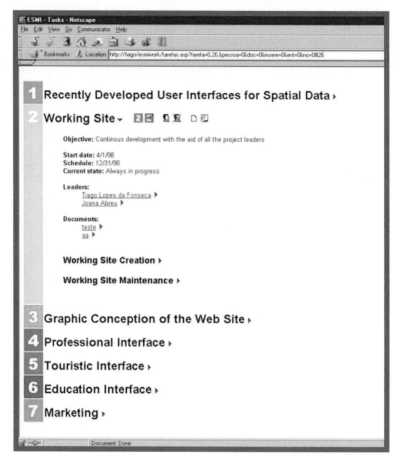

Fig. 3.9 Display of a dynamically generated page

3.3 Geographical Information Systems

3.3.1 Introductory Concepts

Geographical information systems (GIS) are applications that enable the digitising, management, manipulation, analysis, modelling, and visualisation of geo-referenced data. Geographical information systems have a wide range of applications in environmental management such as:

- urban and regional planning;
- natural resource management;
- environmental impact assessment;
- routing and location problems;

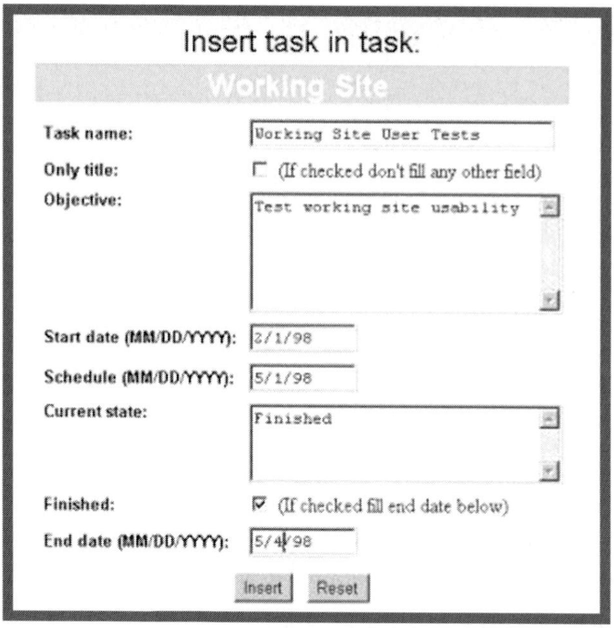

Insert task in task:

Working Site

Task name: | Working Site User Tests

Only title: ☐ (If checked don't fill any other field)

Objective: | Test working site usability

Start date (MM/DD/YYYY): 3/1/98

Schedule (MM/DD/YYYY): 5/1/98

Current state: | Finished

Finished: ☑ (If checked fill end date below)

End date (MM/DD/YYYY): 5/4/98

Insert | Reset

Fig. 3.10 An example of a form for inserting data

```
<%
name = cint(request("name"))
description = cint(request("description"))
.
.
.
Set dbConnection = Server.CreateObject("ADODB.Connection")
dbConnection.Open "Taskjs", "sa"

SQL = "INSERT INTO task ( NAME, DESCRIPTION, START_DATE, END_DATE) VALUE"
SQL = SQL & "('" & name & "','" & description & "','" & start_date & "','" &
end_date & "')"
dbConnection.Execute(SQL)
%>
```

Fig. 3.11 The code reflecting submission of the form

- emergency plans;
- maintenance plans.

Illustrative examples of these GIS applications may be found in Laurini and Thompson (1992), Worboys (1995), Jones (1997), Burrough and McDonnell (1998), and Longley *et al.* (1999).

GIS rely on database capabilities based on special spatial data structures and relational or object-oriented databases coupled with visualisation and analysis facilities. The visualisation includes mapping and additional two- and three-dimensional visualisation tools of the terrain and database attributes. Analysis

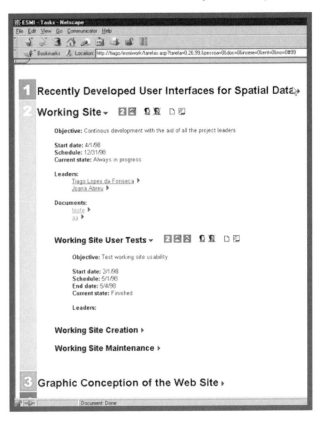

Fig. 3.12 The updated page for the ESMI working site

depends on the type of data model involved, which is discussed later. The development and application stages of a GIS are summarised in Table 3.1.

3.3.2 Terrain Representations

Maps have been the traditional visualisation tools of GIS. They are a symbolic representation of reality with their occasional overrepresentation of features (such as bridges), simplification of irregular lines (such as coastlines), and the use of symbols to represent selected features. Maps are usually of two types: topographic, showing the shape of the terrain; and thematic, displaying concepts such as land use, climate, population densities, and other variables.

Terrains may be seen as a set of spatial entities, which include (Jones, 1997):

- *Point objects*, defined by coordinates (x,y) or (x,y,z).
- *Line objects*, an ordered list of coordinates, that can also be seen as a representation of a mathematical function and may include line segments (finite lines) and half lines (starting at one point but not having an end).

Table 3.1 Development and applications of a GIS

Data acquisition
    Obtain existing datasets (maps, DEMs, aerial photographs, satellite images, alphanumerical information)
    Digitise spatial and alphanumerical information if not available

Preliminary data processing
    Interpret/classify data
    Structure digital data for chosen spatial model
    Transform to common coordinate system

Database construction
    Conceptual data modelling
    Specify database structure
    Specify update procedures
    Load database

Retrieval and analysis
    Retrieve data by location
    Retrieve data by class or attribute
    Find most suitable locations according to criteria
    Search for patterns, associations, routes, and interactions
    Modelling and simulation of spatial phenomena

Communications/visualisation
    Create maps
    Explore data
    Create 3-dimensional views

Source: Adapted from Jones, 1977.

- *Area*. A definition is that a line delimits an area, where the first point equals the last. It can also be seen as a set of lines if an area has holes.
- *Surface* can be seen as: a matrix of points, a triangulated set of points, such as triangular irregular networks of digital elevation models (DEMs), representations for mathematical functions, or contour lines (see Chapter 5).
- *Volume*: a set of surfaces.

DEMs have become increasingly relevant for modelling and visualisation purposes. Modelling of hydrological phenomena (Moore *et al.*, 1991; Mitasova *et al.*, 1995) and forest fires (Gonçalves and Diogo, 1994) are increasingly relying on DEMs to provide a realistic representation of the terrain. DEMs draped with aerial photographs and satellite images are used in visualisation exercises, as reported in Chapter 5.

    Chapter 2 introduced two methods to generate elevation data based on remote-sensed data but there are other methods based on surface-specific point

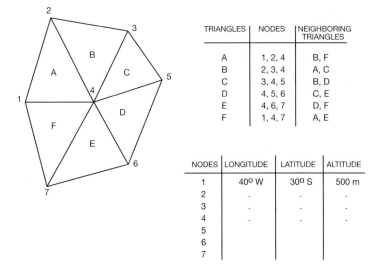

TRIANGLES	NODES	NEIGHBORING TRIANGLES
A	1, 2, 4	B, F
B	2, 3, 4	A, C
C	3, 4, 5	B, D
D	4, 5, 6	C, E
E	4, 6, 7	D, F
F	1, 4, 7	A, E

NODES	LONGITUDE	LATITUDE	ALTITUDE
1	40° W	30° S	500 m
2	.	.	.
3	.	.	.
4	.	.	.
5			
6			
7			

Fig. 3.13 Development of triangular irregular networks

elevation data and contour and stream network data (Hutchinson and Gallant, 1999).

In a DEM, only some points have precise elevations; all the remaining are interpolated. There are usually errors associated with the interpolation, which have severe consequences in applications such as hydrological modelling. To minimise these errors, which tend to create spurious pits and peaks, there have been several proposals reviewed by Mitas and Mitasova (1999). They include local neighbourhood, geostatistical, and variational approaches.

The use of triangulations is a widely applied local neighbourhood approach. It builds topology by unstructured points. This is achieved by triangulating the points as shown in Fig. 3.13, using a process called Delaunay triangulation. This process develops triangular irregular networks (TINs) where:

- the edges have the shortest possible length; and
- the angle between the edges is as large as possible.

The construction of TINs depends on the existence of data points. The most appropriate points are those located at maxima and minima and breaks of slope (Jones, 1997). TINs are used to estimate values at unsampled locations. Linear interpolation methods use planar facets that are fitted to each triangle. Although the results are rapid, they are some of the least accurate methods (Mitas and Mitasova, 1999). Accuracy is particularly important if DEMs are to be used for modelling. It is less relevant if they are to be applied in visualisation exercises.

Geostatistical methods relying on kriging have also been used. However, when local geometry and smoothness are key issues, these approaches may not be adequate (Mitas and Mitasova, 1999). In these cases, the application of functions that pass through the data points and are as smooth as possible may be the best option. These functions are splines (see Appendix 3). Several types of splines have been proposed to improve the representation of terrains, as discussed by Mitas and Mitasova (1999).

Terrains may be characterised by properties such as length, surface area, volume, shape, orientation, and slope (Laurini and Thompson, 1992). However, some spatial properties occurring in terrains cannot be defined in terms of units. They can only be expressed as sets of instances. Examples include (Laurini and Thompson, 1992):

- patterns (clustered or dispersed);
- layout (compact or scattered);
- distances (individual, accumulated);
- enclosures (neighbours, clearings);
- connections and flows;
- sequences (i.e., highways, land use sequences).

3.3.3 Raster and Vector Data Models

A raster model is based on the assumption that space can be divided into cells of a grid (or point objects or pixels). A raster may be mapped onto a matrix. The elements of this matrix are geo-referenced by their Euclidean coordinates. A satellite image or an aerial photograph is representative of this model. Geographical information systems (GIS) include layers of thematic information. For each theme, each cell of the matrix will assume a numerical value. These values may be obtained through sampling or by interpolation.

Raster models are often used on overlay analysis in land use suitability studies. These studies may consist of locating facilities while minimising environmental impact. Overlay analysis relies on operations with the matrices representing the different themes. A composite map (matrix) is obtained by performing operations with source maps (matrices). These operations include the use of Boolean (AND, OR, NOT, and the exclusive OR). They can be applied alone or combined with relational (greater than, less than, equal, greater than or equal to, less than or equal to, not equal) and arithmetic operators (addition, subtraction, multiplication, division, exponentiation, and weighted means).

The ratioing operations with remote sensing images (Chapter 2) are typical raster analytical tools. The map algebra application shown in Fig. 3.19 is another illustrative example. Church (1999) reviews location studies of facilities and corridors using overlay analysis. Examples include siting power plants, waste

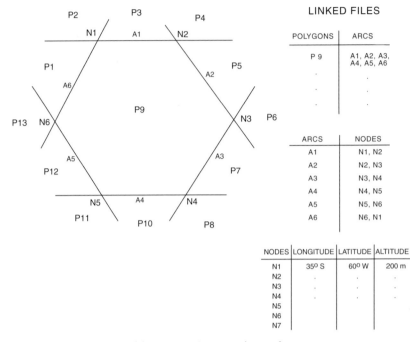

Fig. 3.14 Tables preserving topology of a vector map

water treatment plants, solid waste treatment plants, landfills, highways, pipelines, and power lines.

The vector model represents phenomena using the spatial primitives referred to above (points, lines, areas, surfaces, and volumes). These geometric primitives may have associated non-spatial attributes such as social and environmental variables. Vector models enable the definition of topological relationships. Topological primitives include nodes, arcs (an arc results from the connection between two nodes); directed arcs; rings (close circuit of arcs); polygons or faces (regions bounded by one or more rings); and polyhedrons (regions bounded by polygons) (Jones, 1997). Figure 3.14 illustrates how topology is preserved in the development of tables for a relational database that may be used in a GIS application. These tables enable the use of queries based on the relationships of connectivity and adjacency and the use of network analysis tools. Zhan (1998) describes three relevant classes of network analysis that can be potentially applied in GIS:

- definition of paths, subnetworks, and circuits;
- location-allocation problems;
- network flow problems.

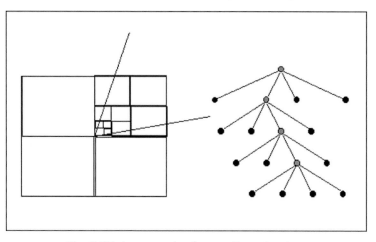

Fig. 3.15 An example of a quadtree structure

Associated algorithms have been applied to a wide number of environmental problems such as:

- solid waste collection routing as proposed in Bodin *et al.* (1989) and Chang *et al.* (1997);
- location of regional waste water treatment systems as discussed by DeMelo and Camara (1994);
- sewer design implemented by Greene *et al.* (1999).

3.3.4 Spatial Data Structures

Spatial data structures have been developed for raster maps or images (Samet, 1989*a,b*). These structures facilitate the storage and retrieval of geographical representations of a same site at different resolutions. The representation of variable spatial resolution within the same scale enables the exploration of different levels of detail of the same area as shown by Muchaxo *et al.* (1999) (see also Fig. 5.33). These structures are also used in map retrieval in services such as CitySearch (see Section 3.6).

The tree data structure is a useful device for storing information required at each level. Each 'node' represents a data item, and its 'branches' are links to other data items. If the process involves the systematic splitting of space in two-dimensional space by a rule of four, then the structure is known as a quadtree (Fig. 3.15). An octree is a three-dimensional equivalent. Another useful structure to store image data is the R-tree. These handle rectangular regions of an image or map. Each R-tree has an associated order, which is an integer K. Each non-leaf R-tree node contains a set of at most K rectangles and at least K/2 rectangles. Each non-leaf tree node must be at least half full. This feature makes R-trees appropriate for disk-based retrieval because each disk access brings back a page including at least $K/2$ rectangles. A disadvantage of R-trees is that the bounding rectangles associated with different nodes may overlap.

3.3.5 Geographical Information Systems and the WWW

The Internet is influencing geographical information system (GIS) development because the Web is considered to be the infrastructure for delivering GIS services to users. The results of this development can be separated into three groups (Alameh, 1998):

- The rise of number of products that traditional GIS vendors are offering on the Internet.
- The increasing demand for interoperability between heterogeneous geographical data types and between geographical and other non-geographical data types, such as multimedia data, which has motivated the Open GIS Consortium.
- The shift towards a service-oriented model for GIS. Users use only a subset of GIS functions as needed, instead of having to buy, install, and learn the proprietary GIS packages. Data and applications will be increasingly delivered as components. Interoperability is a prerequisite for this service model, as users will need to work with heterogeneous data sources—inovaGIS (see Box 3.2), an open source code collection—(see Web site on Open Source movement in Section 3.6) of components follows this model.

Currently, the publication of geographical information on the WWW includes the following alternatives:

- publication of static maps;
- publication and exploration of interactive maps.

The publication of static maps may be accomplished by:

- using the standard graphic support provided by HTML;
- applying image maps, where one may have sensitive areas;
- using tools that provide support for vector formats and added navigation capabilities;
- developing Java applets;
- writing Dynamic HTML in association with executable languages.

The use of tools that provide support for vector formats is the most interesting in the publication of static maps. Vector formats enable unlimited magnification and organisation of the information in different layers for selective visualisation and superimposition. They also require less bandwidth. The drawback is that there is no standard vector format as yet. However, there are alternatives, such as plug-ins from vendors; and the use of macromedia Flash which is becoming a *de facto* standard.

There are a number of Java based applications for geographical information such as ESRI shape files. Portugal Interactivo, a prototype application for the National Centre for Geographic Information also used Java to enable navigation (zoom, panning) and develop information layers including graphic elements vectored on top of aerial photographs and associated alphanumerical information (Fernandes *et al.*, 1997).

The interactive publication of spatial data is possible by using servers that dynamically publish maps to satisfy user requests. Spatial databases become Web-enabled by using a client–server model. Intergraph's Geomedia, Autodesk's MapGuide, ESRI Internet Map Server, MapInfo Map X, and GRASSlinks use this solution.

The possibility of simultaneously querying several Web based geographical information systems has been also researched. A Java based prototype was proposed by Wang and Jusoh (1999) to, among other functions, download datasets and use other software to display or analyse them.

Box 3.2 *inovaGIS*

---

inovaGIS (Gonçalves, 1998) is a component based structure for geographical information implemented as a set of objects. These objects enable querying, map algebra, and visualisation procedures within Windows applications either through the use of Visual Basic or scripting languages, such as JavaScript and Visual Basic Script, among others.

inovaGIS is open-source software freely available at the site referred to in Section 3.6). There are two main modes of using inovaGIS: local processing (Fig. 3.16) or as a remote application (Fig. 3.17).

In the local mode, the user can have access to inovaGIS objects by the way of macros written with scripting languages. He/she can have access to local data or remote data as shown in Fig. 3.17.

For the remote application of inovaGIS, a Geographic Information Remote Server called GIServer was developed. The GIServer is a set of Active Server Pages to access geographical data and functions on the WWW. It uses the inovaGIS ActiveX server in combination with the Microsoft Internet Information Server 4.0 to generate dynamic pages. These behave like a desktop GIS. Figure 3.18 illustrates the concept.

All the geographical information functions are processed in the server but the geographical data may be located in different remote locations. Remote databases can be accessed through SQL queries (structured data), and HTML links related to the subject can also be viewed (non-structured data). The user accesses the data and retrieves HTML pages as text, numbers, and pictures (JPEG and GIF format).

The inovaGIS toolkit includes tools such as an image browser, map algebra, and spatial analysis functions and visualisation in VRML (see Chapter 5). Spatial simulation models based on cellular automata (see Chapter 4) and network analysis methods relying on genetic algorithms are under development.

The image browser enables one to navigate on images larger than 100 megabytes. Figure 3.18 shows the selection of a subset of a large image for display at a much higher resolution.

A classic IDRISI tutorial was selected to illustrate the map algebra possibilities of inovaGIS in Microsoft Excel: knowing the relationship between temperature and topographic data from a regression equation, and having the topographic data, inovaGIS enables the production of a temperature map. Figure 3.19 shows the Excel screen.

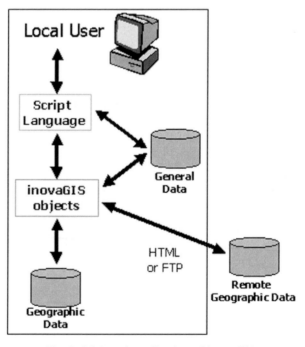

Fig. 3.16 Local application of inovaGIS
Source: http://www.inovagis.org

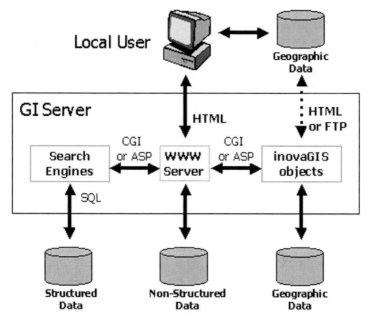

Fig. 3.17 Remote application of inovaGIS

Source: http://www.inovagis.org

10 000 cols x 10 000 rows
Image

Working with a Subset

Fig. 3.18 Browsing an image in inovaGIS

Source: http://www.inovagis.org

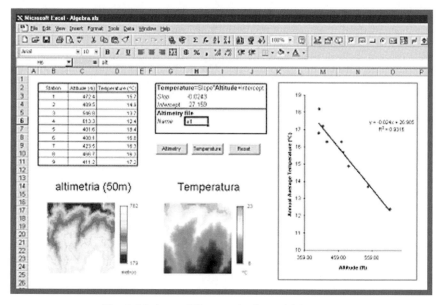

Fig. 3.19 inovaGIS map algebra within Excel

Source: http://www.inovagis.org

The Scalable Vector Graphics (SVG) Working Group of the W3C has evaluated and approved SVG, a language for describing two-dimensional vector, image, and text graphics in Extensible Markup Language (XML). The evaluation included two key criteria: interoperability and scalability. Interoperability requires that vector graphics formats should be supported across different computers, display resolutions, available colours, and data transmission speeds. Scalability means that graphic data can be resised for different display and printer resolutions.

Gould and Ribalaygua (1999) discuss the SVG proposals as related to the Web publication of geographical information. More recently, they have proposed Imapper, a SVG map server (see the Web address in Section 3.6).

3.4 Multimedia Information Systems

Multimedia information systems will be increasingly applied to environmental systems as related imagery and sound data become ubiquitous.

The goal in multimedia information systems is to access data (image, video, sound, and text) and merge their results. Multimedia data, known as metadata, are used to facilitate access. However, multimedia metadata cannot be collected manually for large documents, which makes automatic metadata generation a requirement. In general, multimedia metadata can be classified in three major groups (Klas and Sheth, 1998):

- *Content-dependent*, such as texture and position of objects in an image; individual frame characteristics, such as colour histograms for video objects; and text strings for text.
- *Content-descriptive*, such as the name of authors and year of publication.
- *Content-independent*, metadata that describes the characteristics of the media objects but which cannot be generated automatically, such as the mood reflected by a facial expression.

3.4.1 Image Data

Microsoft Terra Server, University of California at Berkeley's digital library, MIT's orthophoto collection, and SNIG's collection of aerial photos are examples of large databases of image data that can be used in environmental management. Image data can be viewed as collections of objects that are relevant for a given application. Shape and property descriptors define such objects. The former describe the shape of the region within which the object is located. The latter describe the properties of the individual (or groups of) pixels in the given image, such as their red–green–blue values (RGB).

Compression

On an image database it is usually not feasible to store properties on a pixel-by-pixel basis, as images are usually very large objects. Image compression is thus necessary. Compression techniques profit from the redundancies that occur in images, which are of two types:

- *Spatial redundancy*, when neighbouring pixels have the same colour and the same intensity value.
- *Spectral redundancy*, when neighbouring pixels have the same light intensity and colour.

Most images on the WWW which have been used for environmental applications follow JPEG (for Joint Photographic Experts Group) compression. This is the case of aerial photographs available on the WWW. JPEG is a multiple pass compression that includes (Brice, 1997):

- colour space conversion;
- discrete cosine transform;
- quantisation;
- running length and Huffman coding.

Colour space conversion is a transformation that enables the separation of brightness information (luminance) from colour information (chrominance). This transformation calculates Y (representing luminance) from the red (R), green (G), and blue (B) signals, using the expression:

$$Y = 0.3\,R + 0.59\,G + 0.11\,B$$

The Y signal is calculated in such a way that matches as closely as possible the signals from a black-and-white camera scanning the same scene (Brice, 1997). To calculate the amount of colour signals existing in the image that differs from its black-and-white counterpart, one determines:

$$U = m(B - Y)$$

$$V = n(R - Y)$$

U and V are named colour difference signals. As the eye is much more sensitive to luminance, one can use this separation between brightness and colour information to downsample the colour component. This is achieved in JPEG by coding only one U and V colour difference value for every two (block of 2×1), four (2×2), or even 16 (4×4) luminance pixels.

The next step in JPEG compression is to transform the image data to the frequency domain. This is done by applying the discrete cosine transform (DCT) (see Appendix 2 for an introduction to Fourier transforms, FTs; DCTs use cosines instead of the sinusoidal functions associated with FTs) to blocks of 8×8 pixels. This application is performed separately for the luminance and the two-colour difference signals arrays.

With quantisation, the DCT coefficients are serialised by scanning the coefficient array in a zigzag pattern starting at the top left corner and ending at the lower right corner. In quantisation, a pattern matching procedure is followed (Tannenbaum, 1998). In this procedure, each vector or matrix of the original data is matched to a standard pattern. The best match according to some predefined distortion is selected from among a dictionary of possible standard patterns. Then, only the index of the standard pattern is stored.

Because so many coefficients are equal (many to zero), Huffman coding compresses the data further. Huffman coding sets up a table of instances, each one representing a pattern. Pattern recognition results from data observation. These instances point to longer sequences of data. JPEG compression can achieve ratios of 20:1 without significant distortions of the original image.

GIF (Graphic Interchange Format) images are also omnipresent on the WWW. They apply the Lempel–Ziv–Welch (LZW) algorithm which, like Huffman's, is based on the coding of repeated data chains, or patterns. In informal terms, this compression scheme works well if one has large areas with the same colour. GIFs are limited to 256 colours.

Segmentation

From the standpoint of an image database developer, segmentation of an image is a method that may help to identify its relevant properties. Segmentation enables identification of homogeneous regions and interconnections between regions in an image.

Image Databases

The retrieval of images in a database, may be done by applying one of two approaches:

- *The metric approach*, where distances between objects are checked.
- *The transformation approach*, that is, given two objects, 1 and 2, the level of dissimilarity between objects 1 and 2 is proportional to the minimum cost of transforming object 1 to object 2 and vice versa. Transformation operators can include translation, rotation, and scaling (reduction or magnification).

There are several database models that can be applied to images. They include the relational model, the spatial data structure model, and the object-oriented or object-relational model. The relational model can be applied to images, where each image is defined by three sets of properties:

- *pixel-level properties* (i.e., the red–green–blue of each pixel);
- *object-level properties*, properties related to groups or regions of pixels;
- *image-level properties*, such as the date and place of image capture, type of equipment, author, and other similar information on the image (the image metadata).

In a relational database only object-level and image-level properties are represented as relations. Pixel-level properties are explicit in image itself (Subramanian, 1998).

CoastMap is an information system based on a mosaic of aerial photos related to coastal management in the Netherlands (Romao *et al.*, 1995). Figure 3.20 shows the interface of this system which relies on image metadata for retrieval.

Spatial data structures are based on the representation of image databases as trees such as quadtrees or R-trees (see Section 3.3). R-trees, the favoured structure for this data type, are used to store rectangular regions of an image. To represent an image database one has to (Subramanian, 1998):

- Create a relation with two attributes (image identification and object identification), showing which objects appear on which images.
- Create a R-tree that stores all the rectangles.
- Every rectangle has an associated set of fields that contain information about them.

Many existing image database systems are based on the object-oriented model. Images are considered to be objects that can be subject to operations such as rotation, segmentation, and editing (i.e., change of background colour,

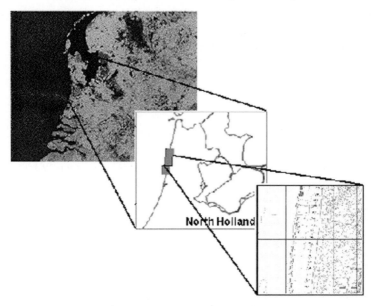

Fig. 3.20 A CoastMap interface
Source: Romao *et al.*, 1995

change of texture, image inversion). For retrieval purposes, current systems rely mainly in the association of a vector of fields (describing image properties such as colour, shape, and texture) to each image. When a user makes a query about an image, he/she creates implicitly a vector that can be compared to the existing database vectors. From this comparison, a set of similar images can be defined and then retrieved by the user.

3.4.2 Video Data

When accessing a video library, one is usually interested in retrieving (Subramanian, 1998):

- *segments* that satisfy given conditions;
- *objects*, i.e., finding predefined objects;
- *activities*;
- *properties*, finding videos or video segments in which objects/activities with certain properties occur.

Compression

Digital video is based on compression that explores not only spatial and spectral redundancy but also temporal redundancy. This redundancy results from the existence of the high number of frames per second associated with video. There are blocks that remain constant from frame to frame and can lead to compression. The quality of the compression method is determined by colour fidelity and number of pixels and frames dropped. MPEG (after Motion Picture Experts Group) is the video compression standard. There are two levels of MPEG:

- MPEG-1 intended for the personal computer.
- MPEG-2 which attempts to provide video at near broadcast quality.

MPEG-1 applies DCT and compression is achieved by quantising the higher spatial frequencies more coarsely than the low spatial frequencies (Brice, 1997), a similar process to the one used in JPEG. However, MPEG also includes audio data.

MPEG-1 videos are stored as a sequence of I, P, and B frames. I represents independent images called intraframes. A P frame is computed from the closest I frame preceding it by interpolation using DCT. A P frame can also be computed from the closest P frame preceding it. B frames are computed by interpolating from the two closest P or I frames. I frames should be decoded first, then P frames, followed by B frames. Once a frame is decoded it can be displayed, while the next frame determined by interpolation is being computed.

MPEG-2 has a higher pixel resolution and data rate. It has higher bandwidth requirements as a result. MPEG-3 has been created for digital television and has even higher requirements.

MPEG video compression can achieve between 50:1 and 200:1 compression rates. MPEG audio compression achieves ratios between 5:1 and 10:1. For a discussion on compression techniques and standards visit the Web site listed in Section 3.6.

Segmentation

Videos are created by taking a series of shots that are then composed by operations such as:

- Shot concatenation.
- Translation, i.e., two successive shots are overplayed one on top of the other.
- Chromatic compositions such as fades and dissolves.

Video segmentation techniques attempt to take a video and determine when shots have been composed using these techniques.

Video Databases

The leading database vendors include video objects in their object-relational systems. A related product relevant for the Internet environment is RealVideo and RealAudio (see Section 3.6 for the Web address). Real systems allow for the creation of both audio and video data and their delivery over the network in a streamed fashion. Streaming is the name given to the technique where a client downloads a portion of the file, decompresses that portion, and starts playing the contents (audio/video) before the rest of the file arrives.

WebSeek (Chang *et al.*, 1997), a system developed to retrieve image and video data on the Internet, uses textual descriptions as well as visual information (see the Internet addresses of the system's Web site and technical report in Section 3.6). It has reviewed more than 600,000 images and videos, many related to nature or environmental problems. Two illustrative screen shots of WebSeek are shown in Fig. 3.21.

Nobre (1999) has developed a spatial indexing system for video. This system is based on the consideration that a given space can be covered with a mosaic of videos. Each of these videos scans a limited area that can be geo-referenced by coupling the camera to a GPS system. From a given point, video sequences are obtained. They cover sections of an urban or rural landscape (Fig. 3.22). Each section is divided in geo-referenced frames. It is possible to retrieve video images by pointing to each frame.

Figure 3.23a shows the proposed interface for this spatial video information system relying on space and time sliders, simple video functions, and sketching facilities to facilitate access. Figure 3.23b displays a related spatial video information system developed by Romao *et al.* (1998).

3.4.3 Audio Data

Audio databases rely on metadata indexing schemes. However, it is sometimes difficult to create metadata and one has to rely on signal processing. Audio databases may be indexed using an audio signal by (Subramanian, 1998):

- *Segmentation*: this is done by splitting up the audio signal into relatively homogeneous windows.
- *Feature extraction*: once one obtains a sequence of windows, k features are extracted from each window. Relevant features include:
  - intensity, the power of the signal generated by the wave;
  - loudness, which does not increase linearly with intensity (see Chapter 2);
  - pitch, which is computed as a derived quantity from frequency and amplitude of the signal;
  - brightness, which is a measure of how clear the sound is.

It is possible to capture audio content through discrete transformations reducing the number of windows. This is done by applying either the discrete Fourier transform (DFT) or the discrete cosine transform (DCT) (see Appendix 2).

Both Informix and DB2 have audio extensions. Informix relies on the MuscleFish audio management system, which uses content based retrievals and has an audio browser. DB2 enables the storing of voice messages from an answering machine.

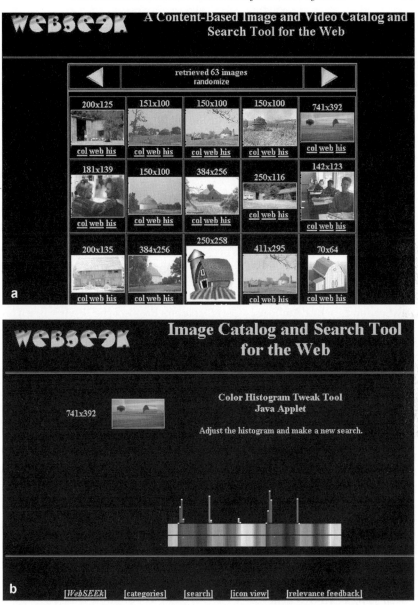

Fig. 3.21 Two illustrative screens of WebSeek: (a) retrieval of images based on a query; (b) Java applet to adjust a histogram and make new searches

Source: http://www.ctr.columbia.edu/webseek. Reprinted with permission

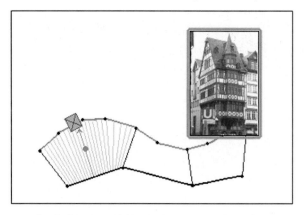

Fig. 3.22 A spatial indexing system for videos

Internet Audio

On the WWW there are two competing technologies: RealAudio and VocalTec (see Section 2.7). RealAudio uses User Datagram Protocol (UDP). It is based on interleaving, whereby if a packet is lost, only one millisecond of data is lost every four milliseconds, rather than a continuous gap of four milliseconds.

VocalTec uses Transmission Control Protocol (TCP). It has an advantage because of the flow control built-in mechanisms, ensuring that the network does not get inundated with multimedia data. Playing the multimedia stream of TCP packets may be interrupted when a packet is lost or delayed.

UDP is faster than TCP. There are packets lost in the transmission. VocalTec uses a predictive caching algorithm that attempts to guess the contents of lost packets (Naik, 1998).

MIDI

Musical Instrument Digital Interface (MIDI) is a standard interface between computers and music synthesisers. MIDI is both a hardware standard and a software protocol: it has a mechanical and electrical specification and a data format specification. With MIDI, most orchestral, synthesiser, and drum parts can be played on one conventional piano keyboard with the various voice synthesisers linked using a MIDI system.

The data transmitted on the MIDI link specifies not only what note has been played, but also the velocity with which the key has been pressed, the force used to hold the key down, and the pedal positions. To replicate a performance, one only needs to record the MIDI data rather than the sound signals (Brice, 1997).

(a)

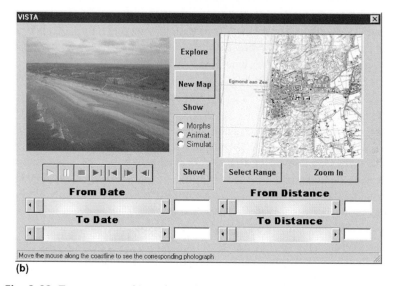

(b)

Fig. 3.23 Two proposed interfaces for spatial video information systems: (a) interface for a frame indexing based system; (b) VISTA, an interface for the repository of videos on the Netherlands' coast

Source of Fig. 3.23b: Romao *et al.*, 1998

3.4.4 Text Data

When discussing multimedia systems, text is overlooked. Text is, however, the media form most widely used in such systems (Tannenbaum, 1998). Text retrieval is hampered for two reasons:

- The same word may have different meanings when used in different contexts (polysemy).
- Different words may mean the same (synonymy).

Metadata for text documents includes content description, storage information, and historical status information (Witten *et al.*, 1994). To identify text documents in response to user queries, searching can be done through full text scans or using document clusters.

Salton *et al.* (1994) presented methods for automatic analysis and search, theme generation, and summarisation of text. These methods also enable the automatic development of links within the text and, thus, hypertext. These methods are based on the representation of segments of text as vectors (sometimes weighted vectors as different terms may have different levels of relevance), and the subsequent comparisons between vectors reflecting text similarity. Commercial databases providing text retrieval include Informix, Oracle (which provides text summarisation), and IBM's DB2.

3.4.5 Multimedia Databases

The leading database software companies offer already extensions to handle image, video, audio, and text. A database specifically designed to handle multimedia data is Mediaway, which has found a niche in marketing applications.

There are also academic efforts to apply Structured Query Language (SQL) to query the Web and extend the search engine capabilities to handle multimedia data, as shown in the sites referred to in Section 3.6. These are limited, however, to multimedia data that are already characterised by appropriate metadata. The only exception is WebSeek, discussed in Section 3.4.2, which automatically creates metadata and enables retrieval based on image or video content. It should be expected that many of the infrastructures discussed below will also develop similar systems to handle the growing stream of image, video, and sound data.

3.5 Environmental Information Infrastructures

Information infrastructures have always existed in the form of libraries, map archives, and paper records. These infrastructures are now becoming increasingly digital as the materials are produced and archived in digital format.

Environmental information infrastructures or more generally, spatial information structures, have been developed around the world. In many ways, they emulate the multinational corporate data warehouse model (Devlin, 1997; Hackney, 1997). However, existing information structures attempt to serve a much wider user base than corporate data warehouses as their organisations are usually public.

Existing spatial information infrastructures rely on meta-information systems that describe a set of spatial databases.

3.5.1 Infrastructure Models

There are two main infrastructure models:

- *The centralised model*, where all the meta-information systems are operated from one location.
- *The distributed model*, where information producers maintain their own meta-information systems but there is a coordinating Web site.

3.5.2 Meta-information

Meta-information takes into account the type of information (alphanumerical, cartographic, remote sensing, other multimedia data), spatial, temporal and thematic referencing, quality, and access. Metadata helps to filter information, overcomes the heterogeneity of databases, and even qualifies uncertainty associated with the information. Metadata includes items such as (adapted from Jones, 1997):

- data exchange format;
- data summary;
- lineage;
- coordinate system;
- specification of primitive spatial objects;
- feature coding system;
- classification;
- geographical coverage;
- temporal coverage;
- positional accuracy;
- attribute accuracy;
- topological accuracy;
- graphical representation;
- an explanation for missing values.

For considerations on more specific environmental metadata see Melton *et al.* (1995).

In Europe, there is a standard for spatial metadata named CEN (after Comité Européen de Normalisation)/TC 287. In the United States, the Federal Geographic Data Committee has proposed a different standard. The reason for these standards is that if one wants to have seamless access to different spatial databases, it would help to have the meta-information associated with each database standardised. If this occurs, distributed models could be implemented without affecting the user.

However, current metadata standards place heavy burdens on data producers, and bottom-up approaches using XML may ease the problem. Data producers could follow examples from Open Source projects (see the Web address in Section 3.6), where developers create document type definition (DTD) and through the emerging exchange with other developers move towards standards (Dougherty, 1999).

3.5.3 Search Mechanisms

The European Spatial Metadata Infrastructure (ESMI) (Geodan, 1997) and the National Spatial Data Infrastructure of the United States follow the protocol developed for digital libraries called Z39.50. This protocol distributed as shareware includes Isite that enables the automatic indexing of query results.

Most systems rely on RDBMS backing of their Web sites and SQL queries. Others use a simple yellow page approach.

3.5.4 National Examples

Europe has several national initiatives. Country examples include (Masser, 1999):

- *Denmark*: the Infodatabase on Geodata is a catalogue that describes the digital maps and other geo-referenced data in Denmark. The meta database gives an overview of each dataset.
- *Finland*, with Map Site, a map service provided by the National Land Survey enables the browsing of maps at different scales. Finland's infrastructure also has a service to send maps electronically to the public and professionals.
- *France*, with the Catalogue des Sources, provides information about public geographical datasets.
- *Germany*, where the Bundesamt für Kartografie und Geodasie maintains ATKIS, a metainformation system for topographic data of the sixteen mapping agencies of the *Länder* (States).
- *Portugal* was a pioneer in Europe in making available in the Internet the National System for Geographical Information (SNIG). Box 3.3 describes the system.

- *The Netherlands*, where the National Clearinghouse for Geographic Information coordinated by RAVI and developed by Geodan, provides information about public and private geographical information.
- *Norway*, which has NGIS, a catalogue of datasets available from the national mapping agency.
- *Slovenia*, through the Geoinformation Centre has a national catalogue of spatial data.

Box 3.3 *Portugal: The National System for Geographical Information (SNIG)*

SNIG was the first infrastructure in the world to make available a catalogue of spatial metadata in the Internet when it opened its Web site in May 1995.

SNIG relies on a metadata database system on geo-referenced information provided by national, regional, and local institutions of Portugal. Metadata is supplied by the data producers following the CEN standard.

SNIG includes map and text based query systems. The infrastructure enables one to finally access the data, either on a free basis or by allowing the data producer to sell it.

SNIG's site map (Fig. 3.24) reflects its comprehensive spatial information. It includes cartographic, aerial photos, satellite images, GPS data, and alphanumerical data. It also shows its role as a portal site for geographical information in Portugal including a library, front-ends for GIS vendors and data producers and community building resources (i.e., Forum SNIG).

Aerial photos and orthophotos covering most of the country are becoming the site's magnet information (Fig. 3.25). Other interesting features under development include SNIG's Excel based mapping tools, its program for schools, on-line repository of urban plans (Fig. 3.26), and fire risk maps (Fig. 3.27).

SNIG started with a professional orientation. Its first site was based mainly on a list of metadata catalogues describing the geographical information produced in Portugal.

The widespread access to the Internet and the relevance of spatial and other environmental data for the general public has inspired the development of the Geocid initiative (Fig. 3.28). Interactive Portugal is an experimental project integrated under that initiative (Fig. 3.29). It includes geo-referenced ground photos and an interface based on sketching (see Chapter 6).

Experimental projects include a geo-referenced complaint system for citizens and VRML fly-over mosaics of aerial photos draping digital terrain models. The latter project is inspired on Digital Portugal, a project discussed in Chapter 5.

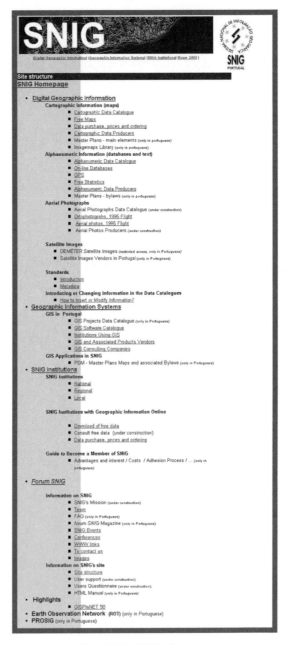

Fig. 3.24 SNIG site map

Image courtesy of the National Centre for Geographical Information (CNIG); SNIG's Web
site (http://snig.cnig.pt.)

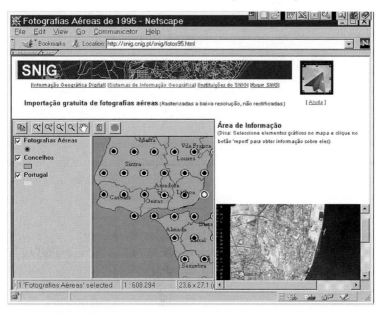

Fig. 3.25 SNIG interface for retrieving aerial photos

Image courtesy of the National Centre for Geographical Information (CNIG);
SNIG's Web site (http://snig.cnig.pt.)

- *Sweden*, where the National Land Survey has the responsibility for the National Metadata Catalogue which includes 150 dataset descriptions from nearly 40 organisations.
- *United Kingdom*, where under the National Geospatial Data Framework, metadata guidelines have been prepared to ensure that any data resources related to the earth's surface are documented consistently.

The United States appointed the Federal Geographic Data Committee in 1990. This committee enabled the development of the National Spatial Data Infrastructure (NSDI) (FGDC, 1994; Nebert, 1996) which includes meta-information, search, and query tools covering federal and state information nodes. NSDI development has been based on client/server technology, Z39.50 and TCP/IP protocols, and providing access to relational databases.

In the United States there are several other national infrastructure systems relevant for environmental management, such as the Bureau of Census, NASA, NOAA, EPA, and USGS systems, discussed in Chapters 1 and 2. The Alexandria Digital Library project of the University of California at Santa Barbara (Smith, 1996) is another spatial data infrastructure relevant from a methodological standpoint in the handling of maps, aerial photos, and satellite images.

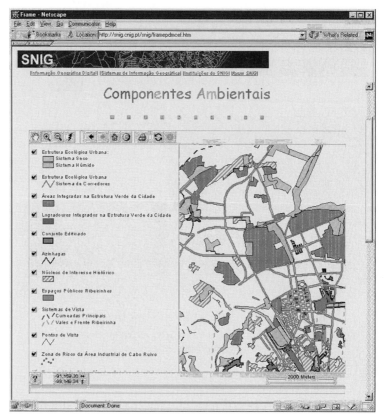

Fig. 3.26 The environmental component of urban plans available at SNIG
Image courtesy of the National Centre for Geographical Information (CNIG); SNIG's
Web site (http://snig.cnig.pt.)

Canadian infrastructures include:

- National Atlas Info Service, enabling the access to cartographic information and to the drawing of simplified maps.
- Canada Centre for Remote Sensing, a meta-information system on remote sensing data.
- Environment Canada, providing access to environmental quality data.

In Oceania, the following infrastructures should be mentioned:

- The Australian New Zealand Land Information Council, provides access to map, aerial photo, and satellite image data.
- ERIN, the Australian Environmental Resources Information Network.

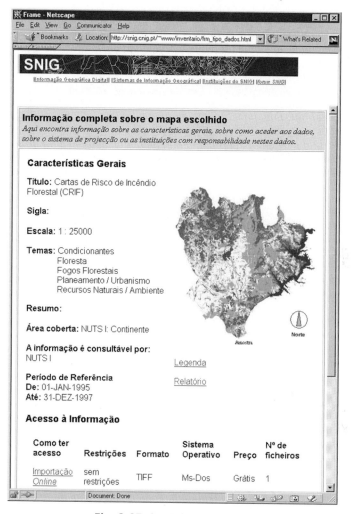

Fig. 3.27 SNIG fire risk maps

Image courtesy of the National Centre for Geographical Information (CNIG); SNIG's
Web site (http://snig.cnig.pt)

3.5.5 Transnational Initiatives

The main transnational initiatives are:

- NASA's Global Change Master Directory, referred to in Chapter 1.
- Megrin Geographical Data Description Directory (GDDD) which contains 200 sets of digital data from 29 countries in Europe.

Fig. 3.28 The Geocid site includes gateways to maps, aerial photos, weather forecasts, and other relevant spatial information for the public

Image courtesy of the National Centre for Geographical Information (CNIG); Geocid (http://geocid-cnig.cnig.pt)

- European Wide Service Exchange (EWSE) developed by the Centre for Earth Observation.
- European Environment Agency (EEA) which coordinates the European Information and Observation Network (EIONET).
- Eurostat which harmonises the statistical data in the European Union.
- EuroGeoSurveys, or GEIX, a metadata server of harmonised geo-scientific information.

In their early stages, environmental information infrastructures were developed with a professional orientation. As shown in SNIG case, and other infrastructures (most notably in the United States, Canada, and Australia) demonstrate, Web sites are also becoming citizen-oriented. This is a reflection not only of the dynamics of the medium itself (WWW), but also the public nature of environmental data.

(a)

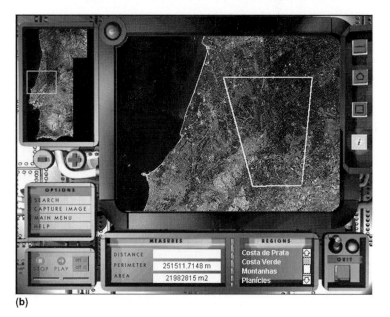

(b)

Fig. 3.29 Interactive Portugal interfaces: (a) ground photo gallery; (b) sketching

Image courtesy of the National Centre for Geographical Information (CNIG)

3.6 Further Exploration

Student projects related to the topics covered in this Chapter may be found at (http://gasa.dcea.fct.unl.pt/camara/index.html). Also at this site, updated information can be found on the sites mentioned below.

Databases

ArsDigita Web site (http://www.arsdigita.com) provides a wealth of information on backing Web sites with database software, including Greenspun's (1999) latest book. Environmental Defense Fund's Scorecard site (http://www. scorecard.org) is an illustrative example. It also offers open source tools to create and maintain Web communities. Leading database vendors include:

- Microsoft (http://microsoft.com) with Access (http://www.microsoft.com/access) and SQL Server (http://www.microsoft.com/SQL).
- Oracle (http://www.oracle.com).
- Informix (http://www.informix.com).
- IBM's DB2 (http://www-4.ibm.com/software/data/db2).
- Sybase (http://www.sybase.com).
- Another alternative database system is Postgres, a non-commercial venture (http://www/postgresql.org).

Database software from these companies is prepared to handle image and video data.

To explore ActiveServer Page information (http://www.ActiveServerPages. com or http://www.15seconds.com). PHP is another server-side, cross-platform HTML embedded scripting language which may be used to develop database backed Web sites (http://www/php.net).

A specialised multimedia information system is Mediaway (http://www. mediaway.com), which has been used by companies around the world to extract marketing material. Webseek is an image and video catalogue, and search tool for the Web (http://disney.ctr.columbia.edu/webseek/about.html).

Many of the leading search engines now enable the retrieval of documents that include multimedia files. Altavista (http://www.altavista.com) is an illustrative example.

An example of an academic system that may also be used to retrieve image and video from the Web is WebSQL developed in Toronto (http://www.cs.toronto. edu/~websql).

Muscle Fish is a leading audio search technology (http://www.musclefish. com).

To understand video and audio technologies in the Web see (http://www.real. com). For audio also visit (http://www.vocaltec.com).

For questions on compression see (http://www.cis.ohio-state.edu/hypertext/faq/usenet/compression-faq/top.html).

Geographical Information Systems

Comprehensive GIS sites include:

- GIS Portal (http://www.gisportal.com).
- Geoplace (http://www.geoplace.com).
- Great GIS Net sites (http://www.hdm.com/gis3.htm).
- GeoInfo Systems site (http://www.gisworld.com).
- National Centre for Geographic Information and Analysis (NCGIA). The NCGIA site provides access to education materials, such as the Core Curriculum on GIS (http://www.ncgia.ucsb.edu/giscc).
- Spatial Odyssey (http://www.odyssey.maine.edu/gisweb/) includes proceedings of leading conferences.

For GIS data resources see:

- Professor Oddens' more than 6600 bookmarks
  (http://oddens.geog.uu.nl.index.html).
- National Geographic's Map Machine
  (http://www.nationalgeographic.com/mapmachine).
- US Census Tiger Files at (http://tiger.census.gov/geo/www/tiger).
- Mapquest provides raster maps for cities worldwide
  (http://www.mapquest.com). Another US map source is CitySearch
  (http://www.citysearch.com), which applies quadtree data structures.
- Other geoportals include CEONet (http://ceonet.ccrs.nrcan.gc.ca), The Geography Network (http://www.geographynetwork.com), GlobeExplorer (http://www.globeexplorer.com), LANDINFO (http://www.landinfo.com), and TerraServer (http://www.terraserver.com).
- Remote sensing sources referred to in Chapter 2.

Information on commercial GIS software is provided at the following sites:

- Environmental Systems Research Institute (ESRI) products
  (http://www.esri.com). ESRI offers distance training from its site.
- Intergraph (http://www.intergraph.com).
- Mapinfo (http://www.mapinfo.com).
- Smallworld (http://www.smallworld.co.uk).
- IDRISI (http://www.clarklabs.org). An on-line tutorial on IDRISI by Eric Lorup may be accessed at
  (http://www.sbg.ac.at/geo/idrisi/wwwtutor/tuthome.htm).

Most of these companies provide WWW services through servers, viewers, and development tools. Autodesk, through its Mapguide product (http://www.mapguide.com) is another example of product migration towards the Internet.

A Vector Markup Language map server is offered by Spain's XYZ (http://www.imapper.com/index.htm). The Internet migration involves interoperability considerations which have been dealt at the OpenGIS Consortium (http://www.opengis.org). This Consortium has inspired the development of GML, a standard way to encode geospatial data in XML (http://www. gmlcentral.com). A general distributed model for distributed geoprocessing in the Internet is provided by Geojava (http://www.geojava.com).

Academic tools for geographical information handling in the Web include:

- Grasslinks (http://www.regis.berkeley.edu/grasslinks), which provides a WWW interface to GRASS (Geographic Resources Analysis Support System) software (http://www.geog.uni-hannover.de/grass).
- Descartes, a Java based visualisation tool (http://allanon.gmd.de/and/and.html).
- A Java based ESRI Shapefile Viewer at (http://www.gis.umn.edu/fornet/java/shpclient).
- inovaGIS covered in this chapter (http://www.inovagis.org), which follows the Open Source model (http://www.opensource.org).

Environmental Information Infrastructures

Major US Federal Agencies which developed information infrastructures for the WWW include NASA (http://www.nasa.gov), EPA (http://www.epa.gov), USGS (http://www.usgs.gov), NOAA (http://www.nooa.gov), and NASA's coordinated Global Change Master Directory. These infrastructures were described in Chapters 1 and 2. USGS hosts the site for the US National Spatial Data Infrastructure (NSDI) (http://www.fgdc.gov/nsdi/nsdi.html).

Canadian infrastructures include Environment Canada (http://www.ec.gc. ca), Canadian Center for Remote Sensing (http://www.ccrs.nrcan.gc.ca/ccrs/), and the National Atlas Info Service (http://ellesmere.ccm.emr.ca).

The Australian Spatial Data Infrastructure is located at (http://www. auslig.gov.au) and Environmental Australia On-line is at (http://www.erin. gov.au).

Sites for infrastructures developed in European countries include:

- Denmark (http://www.geodata-info.dk).
- Finland (http://www.kartta.nls.fi/index.html).
- Portugal (http://snig.cnig.pt). See also (http://geocid-cnig.cnig.pt).
- *Slovenia* (http://www.sigov.si:81/giceng/index.html).
- The Netherlands (http://www.geoplaza.nl).
- United Kingdom (http://www.ngdf.org.uk).

There are several pan-European infrastructures such as:

- Megrin's Geographical Data Description Directory (GDDD) (http://www.eurogeographics.org).
- European Wide Service Exchange (EWSE) for remote sensing data (http://www.infeo.ceo.org).
- European Environment Agency's European Information and Observation Network (http://www.eea.eu.int).
- Eurostat (http://europa.eu.int/en/comm/eurostat/serven/home.htm).
- Geological Electronic Information System (http://eurogeosurveys.brgm.fr/en/geodata.html).
- EUROGI a Pan-European organisation maintains a collection of links to spatial data infrastructures (http://www.eurogi.org).

References

Alameh, N. (1998). *Internet and GIS*. Seminar presented as part of the Planning Support Systems Seminar Series, MIT, Cambridge, MA.

Bodin, L., Fagin, G., and Weledny, R. (1989). The Design of a Computerized Sanitation Vehicle Routing and Scheduling System for the Town of Oyster Ray, New York. *Computers and Operations Research*, 16/1: 45–54.

Brice, R. (1997). *Multimedia and Virtual Reality Engineering*. London: Newnes.

Burrough, P.A. and McDonnell, R.A. (1998). *Principles of Geographical Information Systems for Land Resources Assessment*. Oxford: Oxford University Press.

Chang, N.B, Lu, H.Y., and Wei, Y.L. (1997). GIS Technology for Vehicle Routing and Scheduling in Solid Waste collection system. *Journal of Environmental Engineering*, *ASCE*, 123/5: 901–10.

Chang, S.F., Smith, J.R., Beigi, M., and A.Benitez, A. (1997). Visual Information Retrieval from Large Distributed Online Repositories. *Communications of the ACM*, 40/12: 63–71.

Church, R.L. (1999). Location Modelling and GIS. In P.A. Longley, M.F. Goodchild, D.J. Maguire, and D.W. Rhind (eds), *Geographical Information Systems*, 293–303. New York: Wiley.

DeMelo, J.J. and Camara, A.S. (1994). Models for the Optimization of Regional Wastewater Treatment Systems. *European Journal of Operational Research*, 73/1: 1–16.

Devlin, B. (1997). *Data Warehouse, from Architecture to Implementation*. Reading, MA: Addison-Wesley.

Doughery, D. (1999). XML's Achilles heel. *Web Techniques*, 4/6: 88.

Federal Geographic Data Committee (FGDC) (1994). *The 1994 Plan for the National Spatial Data Infrastructure: Building the Foundations for an Information Based Society*. Reston, VA: Federal Geographic Data Committee, USGS.

Fernandes, J., Fonseca, A., Pereira, L., Faria, A., Figueira, H., Henriques, I., *et al.* (1997). Visualisation and Interaction Tools for Aerial Photograph Mosaics. *Computers & Geosciences*, 23/4: 465–74.

Geodan (1997). *European Spatial Metadata Infrastructure (ESMI)*. Reports submitted to the European Commission, Amsterdam.

Gonçalves, P. and Diogo, P. (1994). Geographic Information Systems and Cellular Automata: a New Approach to Forest Fire Simulation. *Proceedings of the European Conference on Geographical Information Systems (EGIS 94)*, 702–12, Paris.

Gonçalves, P. (1998). *inovaGIS, Tools to Geographic Information Remote Exploration*. Presented at GIS Planet, Lisbon, Portugal (see also http://www.inovagis.org).

Gould, M. and Ribalaygua, A. (1999). A New Breed of Web-Enabled Graphics. *GeoWorld*, 12/3: 46–8 (see also http://www.gisworld.com/gw/1999/0399/399svg.asp).

Greene, R., Agbenowosi, N., and Loganathan, G.V. (1999). GIS Based Approach to Sewer System Design. *Journal of Surveying Engineering, ASCE*, 125/1: 36–57.

Greenspun, P. (1999). *Philip and Alex's Guide to Web Publishing*. San Francisco: Morgan Kauffman.

Gunther, O. (1998). *Environmental Information Systems*. Berlin: Springer Verlag.

Hackney, D. (1997). *Understanding and Implementing Successful Data Marts*. Reading, MA: Addison-Wesley.

Hutchinson, M.F. and Gallant, J.C. (1999). Representation of Terrain. In P.A. Longley, M.F. Goodchild, D.J. Maguire, and D.W. Rhind (eds), *Geographical Information Systems*, 105–24. New York: Wiley.

Jones, C. (1997). *Geographical Information Systems*. London: Longman.

Klas, W. and Sheth, A. (eds) (1998). *Managing Multimedia Data: Using Metadata to Integrate and Apply Digital Data*. New York: McGraw-Hill.

Laurini, R. and Thompson, D. (1992). *Fundamentals of Spatial Information Systems*. London: Academic Press.

Longley, P., Goodchild, M.F., Maguire, D.J., and Rhind, D.W. (1999). *Geographical Information Systems*, Vols 1 and 2. New York: Wiley.

Masser, I. (1999). All Shapes and Sizes: The First Generation of National Spatial data infrastructures. *International Journal of Geographical Information Science*, 13/1: 67–84.

Melton, R.B., De Vaney, D.M., and French, J.C. (eds) (1995). *The Role of Metadata in Managing Large Environmental Science Datasets (Proc. SDM-92)*. Richland, Washington, Pacific Northwest Laboratory, Technical Report No. PNL-SA-26092.

Mitas, L. and Mitasova, H. (1999). Spatial Interpolation. In P.A. Longley, M.F. Goodchild, D.J. Maguire, and D.W. Rhind (eds), *Geographical Information Systems*, 481–92. New York: Wiley.

Mitasova, H., Mitas, L., Brown, W.M., Gerdes, D.P., and Kosinovsky, I. (1995). Modelling Spatially and Temporally Distributed Phenomena: New Methods and Tools for GRASS GIS. *International Journal of GIS*, 9/4: 443–6.

Moore, I.D., Grayson, R.B., and Ladson, A.R. (1991). Digital Terrain Modelling: A Review of Hydrological, Geomorphological and Biological Applications. *Hydrological Processes*, 5: 3–30.

Muchaxo, J., Neves, J.N., and Camara, A.S. (1999). A Real-Time, Level-of-Detail Editable Representation for Phototextured Terrains with Cartographic Coherence. In A.S. Camara and J. Raper (eds), *Spatial Multimedia and Virtual Reality*, 137–46. London: Taylor & Francis.

Naik, D.C. (1998). *Internet Standards and Protocols*. Redmond, WA: Microsoft Press.

Nebert, D. (1996). Draft Implementation Methods for Access to Digital Geospatial Metadata; Normative Annex D; ISO/TC211/WG3/ WI15/N001–US Geological Survey (available at http://www.fgdc.gov).

Nobre, E. (1999). *Spatial Indexing System for Video*. Unpublished report. Environmental Systems Analysis Group, New University of Lisbon, Portugal.

Romao, T., Camara, A.S., Scholten, H., and Molendijk, M. (1995). Coastal Management with Aerial Photograph Mosaics. *Proceedings of the 1st Conference on Spatial Multimedia and Virtual Reality*, Lisbon, Portugal.

Romao, T., Dias, A.E., Camara, A.S., Buurman, J., and Scholten, H. (1998). *New Tools for Coastal Managers*. Presented at GIS Planet, Lisbon, Portugal.

Salton, G., Allan, J., Buckley, C., and Singhal, A. (1994). Automatic Analysis, Theme Generation, and Summarization of Machine Readable Texts. *Science*, 264: 1421–6.

Samet, H. (1989a). *Applications of Spatial Data Structures: Computer Graphics, Image Processing and GIS*. Reading, MA: Addison-Wesley.

Samet, H. (1989b). *The Design and Analysis of Spatial Data Structures*. Reading, MA: Addison-Wesley.

Smith, T.R. (1996). A Digital Library for Geo-Referenced Materials. *IEEE Computer*, 29/5: 54–60.

Subrahmanian, V.S. (1998). *Principles of Multimedia Database Systems*. San Francisco: Morgan Kauffman.

Tannenbaum, R.S. (1998). *Theoretical Foundations of Multimedia*. New York: Computer Science Press.

Wang, F.J. and Jusoh, S. (1999). Integrating Multiple Web-based Geographic Information Systems. *IEEE Multimedia*, 6/1: 49–61.

Witten, I.H., Moffat, A., and Bell, T.C. (1994). *Managing Gigabytes: Compression and Indexing Documents and Images*. New York: Van Nostrand Reinhold.

Worboys, M. (1995). *GIS: A Computing Perspective*. London: Taylor & Francis.

Zhan, F.B. (1998). Representing Networks. Tutorial included in NCGIA Core Curriculum in Geographic Science. Santa Barbara, CA: NCGIA (available at http://www.ncgia.ecsb.edu).

4

Multidimensional Environmental Modelling

4.1 Towards Multidimensional Models

Numerical entities are easily manipulated, interpreted, and have a compact notation (Nickerson, 1988). These features, available data, and limited computing resources have made environmental modelling to be a strictly numerical process until recently.

In the last thirty years, hundreds of numerical environmental models have been developed. Comprehensive reviews may be found in Zannetti (1990) for air pollution modelling, Clark (1996) and Schnoor (1996) for water, air, and soil pollution models, and Chapra (1997) for water quality modelling. The behaviour of ecosystems has also been extensively modelled using both compartment views, such as in Brown and Rothery (1993) and Bossel (1994), and individual-based approaches as in DeAngelis and Gross (1992). Brebbia *et al.* (1997) presented a compilation of noise propagation models and McGuffie (1997) offered a review of climate models.

Most of these models rely on differential equations or partial differential equations solved by numerical methods such as finite differences or finite elements. Some of the models are based on empirical equations. Section 4.6 provides some of the relevant Web addresses for environmental numerical models.

However, numerical models provide limited insights on the workings of real environmental systems. Words, images, and sounds should also be available in such systems. Highly abstract concepts are better represented linguistically as discussed by Rohr (1986). Images (and sounds) provide more realistic views and may illustrate the space and time dimensions of a system.

Beck and Fishwick (1989) discussed the use of natural language in simulation. An earlier application of linguistic simulation models to environmental problems was presented by Camara *et al.* (1987).

Pictorial representations have been increasingly incorporated in the input-output stages of environmental modelling. Examples are the efforts combining environmental modelling and spatial information representations described in the pioneer work of Fedra and Loucks (1985) and Loucks *et al.* (1985), and more recently by Moore *et al.* (1991), Grossman and Eberhardt (1992), Garrote and Bras (1993), Goodchild *et al.* (1993), Mitasova *et al.* (1995), Burrough (1997), Brown *et al.* (1997), Fedra (1997), and Mitas *et al.* (1997). These spatial representations have included maps, satellite, aerial photos, and radar images, and digital terrain models.

There have also been experiments using video to illustrate results of simple water quality models, as described in Camara *et al.* (1993). However, as model results may change dramatically even with small alterations in initial conditions, it is difficult to have the required videos beforehand.

The connection of environmental models to virtual environments has again been explored mainly in the visualisation (and auralisation) of modelling results. Examples include the work of Bryson and Levit (1992), Wheless *et al.* (1996) Chen *et al.* (1997), Gaither *et al.* (1997), Gonçalves *et al.* (1997), and Camara *et al.* (1998), which are discussed in Chapter 5.

Jones (1996) provides an excellent overview on the use of graphics, images, and sound in all the stages of optimisation modelling. His insights can be applied to environmental problems such as water resources (e.g., reservoir operation), solid waste planning, and management (e.g., Chinese postman and travelling salesman algorithms to define routes for waste collection systems), and infrastructure location (e.g., environmental impact assessment of power lines).

Most of these efforts use two or more data types in the modelling effort. However, the modelling engine does not benefit directly from the availability of linguistic, pictorial, or sound data. The core of these models is still a simplified representation of reality expressed in numerical terms.

The availability of multimedia data, such as static imagery and video, the expanding computational resources and developments in fields, such as artificial life (Reynolds, 1982, 1987; Sims, 1994; Taylor and Jefferson, 1994; Terzopoulos *et al.*, 1994), and sensory ecology (Dusenbery, 1992), have opened the possibility of exploring new environmental modelling approaches. These approaches represent an 'agent' view of environmental systems (Flake, 1998). One can study the environment at different levels, but there are always two types of phenomena: agents (molecules, animals) and interactions of agents (chemical reactions, predator–prey relationships). These new methods also intend to be quasi-realistic in their representations of agents and their interactions. They can be divided in four major groups:

- cellular automata;
- other individual based approaches;
- programming by reproduction;
- multidimensional modelling.

All these approaches include state variables (the dependent variables being modelled can be seen as an agent or a collection of agents with the same properties), parameters (coefficients characterising the agent interaction functions) and inputs (forcing functions or constants) such as the more traditional environmental models. This means that methods, such as causal diagramming discussed by Coyle (1983), and calibration (tuning of model parameters), verification of model results and sensitivity and uncertainty analyses, as presented by Schnoor (1996) can be applied with minor modifications.

4.2 Cellular Automata Approaches

The dynamics of environmental systems have been traditionally modelled using three separate approaches as outlined by Grossman and Eberhardt (1992):

- *Dynamic models* using spatially aggregated variables but with results that can be modified, for each site, by locally varying parameters.
- *'Active area dynamic' models*, where the development of each area is modelled separately with inputs from adjacent areas.
- *Transport models* based on physical laws or empirical principles.

Environmental systems may be seen as a set of discrete interacting agents. Traditional partial differential equations, however, start by considering continuum approximations to these discrete systems. The continuum is then discretised to solve the equations using numerical methods.

Toffoli and Margolus (1987) and, more recently, Gershenfeld (1999) have argued that if the goal is to solve the problem with digital computers one should not use continuum descriptions. One can start with discrete representations and simulate them using cellular automata approaches. Cellular automata models of dynamic systems consider a lattice of cells on a vector (one-dimensional cellular automata), or a grid (two- or three-dimensional cellular automata), with each cell containing a few bits of data. Time advances in discrete steps and the cells determine their state at the next time period through transition rules. These rules take into account the values of the cells and their neighbours in the previous time step. Transition rules for environmental models may be derived from physical laws, empirical observations, or expert opinions. Figure 4.1 shows examples of transition rules for cellular automata.

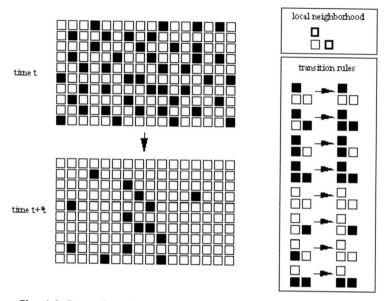

Fig. 4.1 Examples of transition rules for cellular automata models
Source: Castro, 1996

In computational terms, cellular automata are similar to the numerical methods used to solve partial differential equations. The differences are that each cell of the lattice is only allowed a small set of possible values and the transition rules may not be in algebraic form. These rules may be deterministic or stochastic (Hogeweg, 1988). Typical benefits of cellular automata include:

- a model representation closer to the real physical system;
- it does not involve any power-series truncation as in traditional numerical methods;
- it is not subject to round-off;
- it can be easily extended from one to two and three dimensions.

The definition of appropriate transition functions and the implementation of models that have several state variables (i.e., water quality models considering several contaminants) represent often-significant shortcomings of the cellular automata approach.

4.2.1 Cellular Representations and Boundary Conditions

Cellular automata models assume a discretisation of space in mosaics of cells. Traditionally, the mosaics of cells, or tesselations, used in cellular automata models are based on cells with identical size and shape (regular tesselations). Square tesselations present significant advantages due to the equality of sides, decomposability, and stability of orientation and aggregation as pointed out by Laurini and Thompson (1992). They are dominant in grid cell data capture. They are also present in the form of arrays of pixels in an image which makes cellular automata a valuable tool for image based simulations, as considered by Romao *et al.* (1998).

In cellular automata models, one can consider transition rules as a method of propagation or elimination of a finite number of tokens. The behaviour of the transition rules at the boundaries of the model domain may be of three dominant types (Camara *et al.*, 1996):

- *Cyclic*: the tokens crossing a boundary may reappear in a cell at the opposite boundary.
- *Finite*: the tokens are not able to transpose the boundary so they tend to accumulate near it.
- *Infinite*: the entities that cross the boundary disappear from the simulation.

4.2.2 Examples

Cellular automata have been applied to:

- Ecological modelling (Hogeweg, 1988; Phipps, 1992; Camara *et al.*, 1994).
- Surface water quality modelling (Castro, 1996).
- Urban and regional planning (Itami, 1994; Xie, 1996; White and Engelen, 1997).
- Forest fire modelling (Gonçalves and Diogo, 1994; Karfylidis and Thanailakis, 1997).
- Nonpoint pollution modelling (Camara *et al.*, 1998).
- Oil spill modelling (Romao *et al.*, 1998).
- Slothower *et al.* (1996) have discussed their use in connection with raster geographical information systems.

These applications fall into two main categories: individual based ecological models and diffusion models (fire, oil spills, water quality, and land use). For illustrative purposes, a predator–prey and a forest fire model are discussed.

Predator–Prey Model

Camara *et al.* (1994) have modelled a predator–prey relationship using a cellular automata formulation and compared the results with the ones obtained with a traditional Lotka–Volterra model.

Predator and prey were located in cells of a mosaic representing a territory (Fig. 4.2). Random reproduction and death rules were defined using probabil-

Fig. 4.2 A predator–prey cellular automata model

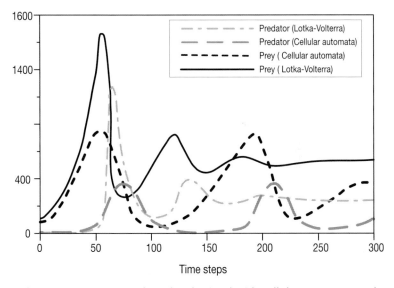

Fig. 4.3 Comparison of results obtained with cellular automata and
Lotka–Volterra approaches

ities derived from the corresponding Lotka–Volterra coefficients. Predator and prey could meet in the same cell. For a certain probability, the prey could die.

Figure 4.3 shows the difference between the results obtained with the two approaches: one that considers the spatial characteristics of the individuals (the cellular automata approach) and another that only takes into account their number. Note that calibration and verification procedures would be identical for both approaches.

Forest Fire Model

Landscapes can be represented as cellular automata, where each cell represents a certain area and has attributes such as topography and vegetation cover. The most widely used forest fire model is Rothermel's (1983), which enables the forecast of the fire's speed of propagation and intensity based on the following parameters:

- fuel characteristics such as moisture;
- proportion of live and dead material;
- density;
- surface volume coefficients and mineral content;
- wind speed and direction;
- altitude and slope.

Gonçalves and Diogo (1994) developed a cellular automata model for forest fire propagation (FireGIS) combining a cellular automata representation of a landscape and Rothermel's model. They made the following assumptions:

- The fire speed of propagation can be input in the cellular automata model by applying the Rothermel model to determine it.
- Vegetation is homogeneous within each cell.
- There are eight independent wind directions.
- Slope is null in each cell.
- Only cells with the fire already extinct or without vegetation are not subject to fire propagation.
- In each time step, each cell can propagate the fire to only one of the eight adjacent cells.

Fire ignition may start in one or more cells. Each cell may then have one of four possible states: unburned; fire with no capacity to propagate; fire with capacity to propagate; and burned. Figure 4.4 shows how the model operates. To evaluate the results of the fire spread model, images of burned areas, as determined by the model, are compared to real images of burned areas. This evaluation may use the model proposed by Turner *et al.* (1989), which includes the calculation of shape and area indices such as fractal dimensions (Peitgen *et al.*, 1992) and the burned areas interface index (Turner *et al.*, 1989).

Gonçalves (1997) has observed the patterns of a fire represented in different scales (Fig. 4.5). He then estimated the fractal dimension of a fire by log plotting (see Fig. 4.6) the number of cells with the presence of fire and the corresponding length scale (the fractal dimension $D = 1 - m$, where m is the slope). It is possible

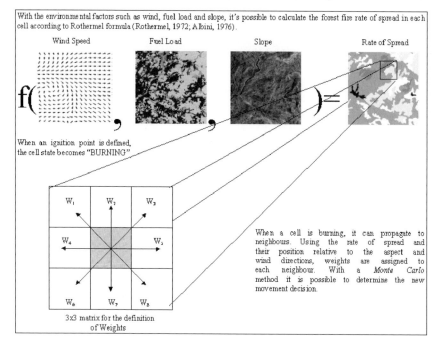

With the environmental factors such as wind, fuel load and slope, it's possible to calculate the forest fire rate of spread in each cell according to Rothermel formula (Rothermel, 1972; Albini, 1976).

Wind Speed　　　　　Fuel Load　　　　　Slope　　　　　Rate of Spread

$f($　　　　　　　　　　　　　　　　　　　　　　$)=$

When an ignition point is defined, the cell state becomes "BURNING".

W₁ | W₂ | W₃
W₄ | | W₅
W₆ | W₇ | W₈

3x3 matrix for the definition of Weights

When a cell is burning, it can propagate to neighbours. Using the rate of spread and their position relative to the aspect and wind directions, weights are assigned to each neighbour. With a *Monte Carlo* method it is possible to determine the new movement decision.

Fig. 4.4 Cellular automata for fire spread modelling
Source: Gonçalves, 1997

Length scale

Fig. 4.5 Forest fire patterns across scales
Source: Gonçalves, 1997

to observe that the geometry across scales is similar by overlaying the patterns (Fig. 4.7).

By running FireGIS for 4, 6, and 12 hours, it was shown that the fractal dimension for the two latest simulation times was closer to the fractal dimension of a real fire (Fig. 4.8). For shorter simulation times, it was found that small-scale features become more relevant than geo-morphological factors. As a result, the fractal dimensions obtained with the model were not similar to those obtained by the analysis of the real fire.

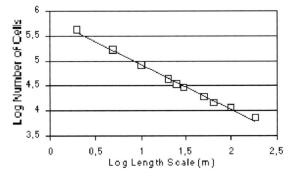

Fig. 4.6 Evaluation of the fractal dimension of a fire

Source: Gonçalves, 1997

Fig. 4.7 Multiresolution fire patterns

Source: Gonçalves, 1997

4.2.3 Implementation Issues

The predator–prey cellular automata example was implemented using Visual Basic. Gonçalves *et al*. (1997) have implemented the forest fire spread model in the inovaGIS environment described in Chapter 3. Within inovaGIS, a generic data object (called iData) has two descendants: iVector and iRaster, to deal with vector and raster data, respectively. An object called iAutomata was developed as a descendant of the raster object. This object enables cellular automata implementations.

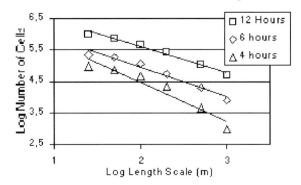

Fig. 4.8 Fractal dimensions obtained with the FireGIS model

Source: Gonçalves, 1997

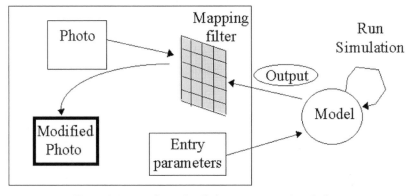

Fig. 4.9 Image based cellular automata simulation

Source: Romao *et al.*, 1998

Benefiting from the inovaGIS interoperable environment, one could use iAutomata to run the model using iRaster data (maps, aerial photos and/or digital terrain models obtained from IDRISI) and have Excel as a front-end to input model parameters and visualise the model results. In addition, one could visualise the model results in a virtual environment, VirtualGISRoom, discussed in Chapter 5 (see Fig. 5.34).

Two other relevant cellular automata implementations include Romao *et al.* (1998) and Castro (1996). Romao *et al.* proposed a cellular automata model running on an aerial photo to produce a changed photo as shown in Fig. 4.9. This process, IMSIM, relies on the application of an image-processing filter (see Appendix 2) that follows cellular automata rules.

Castro (1996) implemented a dynamic cellular water quality model using parallel processors in an iPSC/1 system. The particle tracking based model

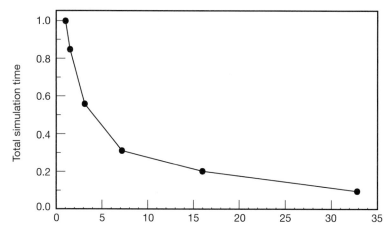

Fig. 4.10 Implementation of a cellular automata model with parallel processing. Even using more than 30 processors, the gains tend to level off rather than increase

Source: Castro, 1996

included advection, diffusion, and decay processes modelled as probabilistic transition rules. The results were comparable to the ones obtained with traditional differential equation based water quality models. It was also shown that it was possible to significantly decrease the simulation time by increasing the number of processors, but that such improvement tended to level off due to overhead communication. In that particular case, after allocating more than thirty processors there were no appreciable benefits (Fig. 4.10).

4.3 Other Individual Based Approaches

Cellular automata provide a framework for individual based models but there are alternative methodologies to model individuals and their behaviours. Craig Reynolds provides a list of annotated Web links (see Section 4.6) on these approaches which are also called entity or agent based simulations. Reynolds (1982) introduced many techniques which have subsequently been applied in multi-agent simulation. In this type of simulation, agents are autonomous units that interact with their environment (including other agents) but act independently. They do not take commands from a leader and do not follow a global plan.

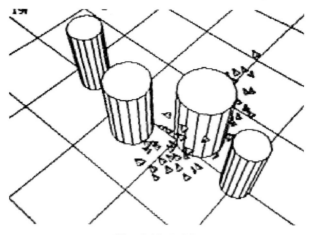

Fig. 4.11 Boids

Source: Reynolds (1987); an animation of boids may be found at http://hmt.com/cwr/ibm.html

In 1987, Reynolds introduced the concept of 'boids' (Reynolds 1987). A Java animation of boid behaviour is shown at his Web page (see Section 4.6). This behaviour follows four basic rules (Fig. 4.11):

- *Avoidance*: move away from boids that are too close, to reduce the chance of collision.
- *Copy*: fly in the general direction that the flock is moving by averaging the other boids velocities and directions.
- *Centre*: minimise exposure to the flock's exterior by moving towards the centre.
- *View*: move away from any boid that blocks the view.

Gonçalves and Antunes (1995) made an effort inspired by Reynolds' work to simulate a flock of birds. Birds were treated as vectorial objects characterised by their location data (x, y, and z coordinates) and an optional orientation vector. To achieve an interconnection between different objects there is a function of attraction working as a weight that expresses its dependence from the neighbourhood, as in weighted Voronoi diagrams (WVD). The Voronoi diagram is a system that provides explicit proximity information about a finite set of points in an Euclidean space. Given a finite set, S, of points one can consider the simplest case as a subdivision of the space into mutually disjointed regions according to the nearest neighbour rule.

Two social behaviours were introduced in this model: attraction, working by means of relatedness and sex, which achieves a grouping process; and repulsion that, by contrast, incorporates aggression, dominance, or simply avoidance

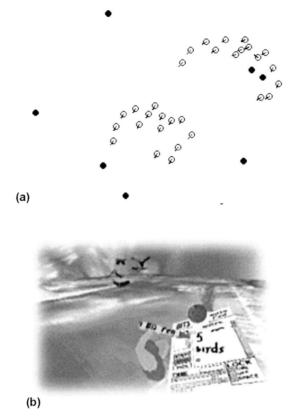

(a)

(b)

Fig. 4.12 (a) Weighted Voronoi based model of a flock of birds;
(b) visualisation on a virtual environment

Source: Gonçalves and Antunes, 1995

between the objects. Figure 4.12 shows the underlying model and the virtual
environment visualisation of a flock of birds. There have been several other
modelling projects of artificial animals, for example:

- Miller (1988), focusing on snakes and worms.
- Terzopoulos *et al.* (1994), modelling artificial fish.

In the latter work, there is a quest for realism not only at the level of superficial
appearance but also at the level of the physical behaviour of the animal within
its environment. These considerations included locomotion, perception of the
world, behaviour, and ability to learn. Other related approaches have been
the simulation of humans, proposed by Granieri and Badler (1994). These
approaches use procedural animation techniques described in Chapter 5.

Gomes (1997) proposed an approach for modelling a school of artificial fish in a virtual environment, based on the work by Terzopoulos *et al.* (1994). A school is a set of autonomous animals that may behave in a fluid-like fashion as if there was a centralised control. However, the evolution of a school is the simple result of actions of several individuals, each one following its perception of the surrounding world. The animals may avoid obstacles and each other and be hunters and prey. The model is highly parameterised, enabling an expert user to configure his/her own ocean, with his/her own fish. One can have as many species as he/she wants with a number of primitive/reflex behaviours (e.g., obstacle avoidance) and more complex behaviours (e.g., hunting, escaping) which depend on a mental state (e.g., hunger, fear).

The artificial fish have an overview of the world, by means of simulated visual perception, in a vision volume. The fish have direct access to a database describing the position, orientation, and exact velocities of all the objects in the environment. However, the information provided to the animal by this database is limited by realistic considerations. For instance, one cannot forget that objects can hide behind other objects. In the model, the animals' neighbourhood was defined as a sensitive volume, centred on the animal. The magnitude of the sensitivity is defined as $1/e^d$, where e is the exponent and d is the radius. The contribution of each animal is $1/d^2$, where d is the distance. In the model, there are five cumulative volumes as shown in Fig. 4.13.

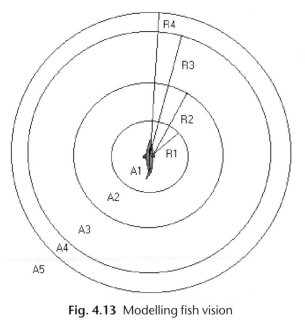

Fig. 4.13 Modelling fish vision

Source: Gomes, 1997

The model includes other parameterised variables, for example:

- *School behaviour*, which means that one can specify it if a given set of fish behave as a school.
- *Mating behaviour*, when one specifies that two fish can mate.
- *Appetite*, where one can specify the need for a predator to hunt.
- *Pursue and escape*, which are related to predators pursuing prey that escape.
- *Calories*, where one specifies the amount of calories represented by an ingested prey.
- *Velocity*, the speed of a fish.

The software used in the development process represents entities geometrically and these behaviours are programmed externally. This means that more complex ecological processes can be added.

There have also been efforts to model evolution using genetic programming, where one searches for individuals with the highest fitness values possible, such as described in Sims (1994). Section 4.6 provides the Web address of a retrospective on Sims' evolving creatures' project. Other relevant artificial life projects include SWARM, The Virtual Fish Tank, and Nerve Garden. SWARM is a software package for multi-agent simulation of complex systems developed at the Santa Fe Institute (see Section 4.6), which has several environmental applications. The Virtual Fish Tank, a development inspired by Resnick's work (1994), is designed for environmental education. Figure 4.14 shows a Virtual Fish tank screen. In the Virtual Fish Tank, users can:

Fig. 4.14 A Virtual Fish Tank exhibit at the Computer Museum in Boston, Massachusetts

Source: Image courtesy of Nearlife (http://www.nearlife.com)

- create their own artificial fish;
- design behaviours for their fish;
- play with their fish;
- observe interactions among fish;
- analyse ecological patterns that emerge.

Nerve Garden is a biologically inspired, collaborative world available on the Internet (address in Section 4.6). It represents a development on earlier work on digital ecosystems (Ray, 1991), evolving creatures (Sims, 1994), and the work on L-systems (Prusinkiewicz and Lindenmayer, 1990). Users can grow plants and submit the data onto a VRML97 scene called the Seeder Garden. A cellular automata engine is being developed to simulate the growth and replication of plants and introduce a class of virtual herbivores (Damer *et al.*, 1999).

Campos and Hill (1999) demonstrated how multi-agent models for ecosystem simulation may be implemented on the WWW using Java and VRML visualisation (see also their Ecosim Web site referred to in Section 4.6). Figure 4.15 shows the interface of the Ecosim system. Many of these efforts represent a trend towards more realistic models of environmental systems. They present two major requirements:

- A demand for knowledge at a micro level.
- The integration of simulation with animation techniques.

Environmental studies deal with real ecosystems, which means that models have to include transition rules representing natural behaviour. Most of the existing models have done so but at a macro level. Concepts from sensory ecology which include animal vision and hearing and their reactions to environmental stress (Dusenbery, 1992) are still not applied in most models. Camara *et al.* (1998) tried to incorporate some of those concepts in modelling the reactions of fish to water pollution in a virtual environment. Using the properties of such an environment, an attempt was made to enable the user to impersonate a fish and acquire its vision as described in Chapter 5. There is a long way to go, however, as sensory ecology is still in its infancy.

Realistic simulations such as those described here usually include a simulation/animation loop proposed by Ventrella (1999). This loop is far more demanding than the traditional numerical modelling method. It includes the following steps:

- User interaction is achieved by a mapping mouse, keyboard, or any other device to cause physical events in the simulated world (i.e., animals and plants can be moved or backgrounds and artificial objects can be modified). The user can also change the viewpoint.
- As a result of user interaction, the effects on the entities of the simulated world have to be computed. These changes may trigger

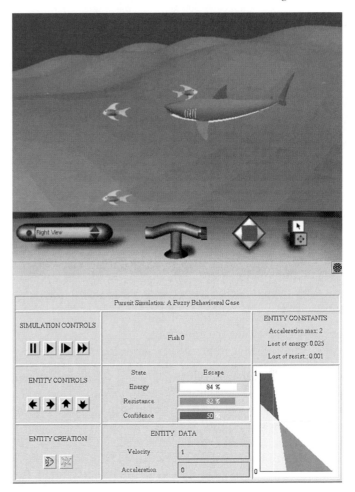

Fig. 4.15 Illustrative screen of the Ecosim system

Source: Campos and Hill, 1999. Copyright 1998, Simulation Councils, Inc., from
Transactions of the Society for Computer Simulation

environmental changes and simulation models may be used to represent
these changes.

- Independent of user input, there are natural evolutionary dynamics that
  have to be computed: growth, reproduction, mutation, and death. There
  are also flows and transfer of energy that need be modelled.
- Physical processes have to be considered. They include gravity, ground
  collisions (e.g., bouncing), and water friction.
- Graphics for each animation frame have to be rendered and displayed.

4.4 Programming by Reproduction

Programming by Demonstration (PbD), where the user demonstrates the behaviour of objects to the computer, has been widely discussed in recent years (Cypher, 1993). PbD simulation based models are similar to other individual based approaches. The major difference is that transition rules are defined by example. One of the better-known simulation applications of PbD is KidSim (Canfield-Smith *et al.*, 1994), where behaviour and interaction rules between objects were demonstrated a priori by users. Other applications have been proposed by Lieberman (1994, 1995, 1998) and Friedrich *et al.* (1996). Animation systems, such as Macromedia Flash, also use demonstration principles.

PbD systems require the user to have sufficient skill and knowledge to program what he/she wants to accomplish. This may be too difficult to achieve even for systems that have a small number of objects. An example is a predator–prey system where the relationships between them (i.e., attraction and repulsion) depend on a number of factors such as distance and number of predators and prey.

Nobre and Camara (1999, 2001) introduced Programming by Reproduction (PbR) inspired by PbD but with much less overload for users. In PbR, behaviour and interaction rules between objects or agents can also be demonstrated; however, they can be reproduced from digital video.

The use of video as a data source for models is the main contribution of PbR. Users only need to recognise the agents and their actions and reproduce them into the computer by sketching directly on the video. Then, a simulation model can be run, which may have quantitative, graphical, and even video outputs. PbR can make use of videos already available and may open new alternatives on the use of video in environmental data sensing.

4.4.1 Information Acquisition

In PbR, spatial and temporal information on objects is extracted by following their movement using sketching. Each point of a line can be labelled with its spatial and temporal position (Fig. 4.16). By combining measures of spatial variation and temporal variation for each point or to a set of continuous points in a sketch, a number of parameters can be inferred, as shown in Table 4.1. To obtain these measures, spatial and temporal rulers are applied. These rulers have to be previously calibrated. For the spatial ruler, the length of an image pixel has to be known and it will set the scale. This ruler has to be updated, whenever the video sequences change their associated scale. The temporal ruler depends on the frame rate that is typically constant for the entire video. Depth perception and perspective factors are the main problems related to the use of digital video in PbR. One way to solve the problem is by changing the scale within each image by using control points as shown in Fig. 4.17.

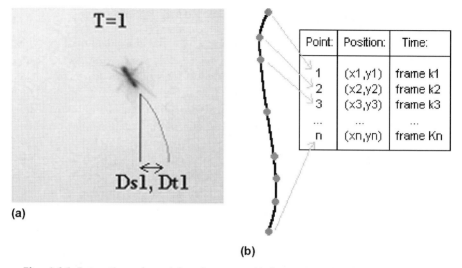

Fig. 4.16 Extraction of spatial and temporal information: (a) sketching on the video; (b) a generated table

4.4.2 Simulation

With the objects and their behaviour and interaction rules defined, a simulation model can be implemented by reproducing those rules. Objects can be represented as iconic cells defined by their position. In a predator–prey model, one could have predator and prey cells. By placing them in a computer screen, the first screen of a cellular automata model is obtained. The whole process is shown in Fig. 4.18. PbR rules can also be developed with Programming by Demonstration (PbD) rules, originally developed by Camara *et al.* (1994). An updated list is shown in Table 4.2.

Objects can be seen as pictorial entities. Until now, in this explanation, they have been treated as iconic entities defined by their position. However, other pictorial variables, such as shape and size, can be considered. The evolution of shape and size involves the definition of a deformation model. In environmental systems, shape deformation may be used to model forest fires and oil spills, for example. Nobre and Camara (1999) have modelled an oil spill using such a technique. In this model, currents and winds are vectors that can be sketched and affect directly the oil spread as shown in Fig. 4.19. The basic oil spill model is a cellular automata formulation and the sketched vectors create vector fields that interact with the cellular automata representation (Fig. 4. 20). (An animation of this model is included on the book's Web Site.)

Triggering winds and currents (or other vector fields) by sketching may also be used to drive particle based cellular automata models as in Fig. 4.21. A shape

Table 4.1 Object parameters inferred by sketching on a video

Primitive variables	Composed variables (measures)	Formulae	Explanation	System variables (some examples)
Space (m)	Local distance	Ds	Distance between 2 continuous points	Space covered in a persecution
	Total distance	Σ(Ds)	Sum of all local distances	
	Minimum distance	$P_{end}-P_{ini}$	Linear distance between the first (Pini) and last point (Pend)	Prey–predator distance detection
Time (s)	Local time	Dt	Time variation between 2 continuous points	Pollutant time discharge into a river
	Total time	Σ(Dt)	Sum of all local time	
	Local velocity	Ds/Dt	Velocity between each 2 points	Wind speed; animal running speed
	Average velocity	Σ(Ds)/Σ(Dt)	Average velocity for the complete stroke	
	Local acceleration	$2(Ds/Dt)/Dt^2$	Acceleration between 2 continuous points	Animal acceleration during a persecution scene
	Average acceleration	2(Σ(Ds)/Σ(Dt))/Σ(Dt)	Average acceleration for the complete stroke	
	Expansion	Expansion rate	Object's area, diameter, or radius increment	Oil spill expansion rate
	Retraction	Retraction rate	Object's area, diameter, or radius decrement	Coast erosion
	Pattern	Direction variation	Object's movement patterns (time variation on the movement, speed, and direction)	Search pattern of animal for food

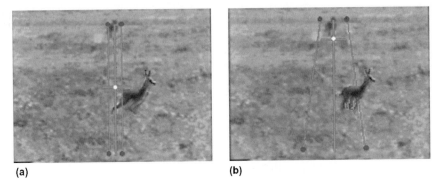

(a)   (b)

Fig. 4.17 Depth and distortion correction on digital video: (a) setting control points; (b) adding perspective

deformation model may also be overlaid on a particle based model for an oil spill as shown in Fig. 4.22. It is then possible to recognise the spill boundaries rapidly, and at the same time have detailed information on relative crude concentration.

4.4.3 PbR Implementation

Video Sketch is a prototype of PbR that includes five main tools (see Fig. 4.18):

- *Video manipulation* to select and control the video flow.
- *Video sketching* to define objects, sketch and take notes on the video images and simultaneously infer dynamic measures for speed, attraction, and repulsion among objects.
- *Object definition*, where these sketches and notes are linked to objects, a relational graph is created, and the dialogue boxes to define behaviour and interaction rules are filled.
- *Simulation*, which includes a basic screen (named scenario) where the objects are placed and can be edited (objects may be added or eliminated at will). In the simulation model, the behaviour and interaction rules observed in the video are reproduced automatically. One can, however, change them at any time during the simulation run.

A background scene can be imported to the scenario and instead of iconic cells, video sequences may be used to represent the animals. These sequences are similar to the animated buttons used in many animation packages. Using this method, the model output can seem to be a new video (Fig. 4.23).

To improve the quality of video output, one has to take into account that video sequences of the animals change as their behaviour changes. This means that the

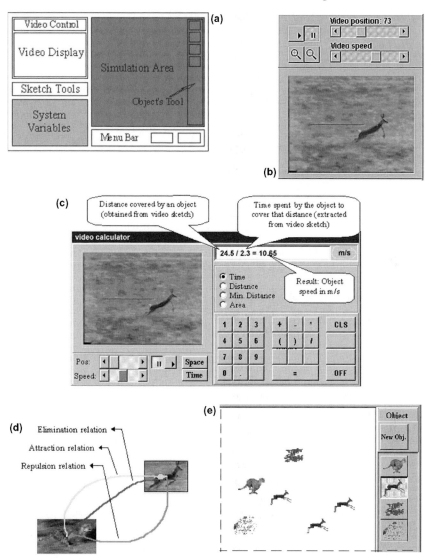

Fig. 4.18 Programming a predator–prey model by reproduction: (a) layout of computer implementation; (b) and (c) information acquisition; (d) causal diagramming; (e) model scenario

video has to be investigated thoroughly and the video sequences classified (which may be helped by supervised image classification methods), stored on a database, and retrieved on call by the simulation model. The video representation of the model results may also become interactive: one may add or delete entities and

Table 4.2 Programming by demonstration rules

PbD rules	Iconic representation	Example
Movement		*Animal movement* Probability of movement and orientation are key parameters
Expansion		*Fire spread* Wind speed and direction dictate probability and orientation of expansion
Decay		*Bacterial decay* Bacteria mortality rate dictate probability of bacterial decay
Death		*Virus* Animal death due to virus
Reproduction		*Animal reproduction* Animal reproduction as a function of animal encounters

change the background and then observe the video output. Using this method, interactivity in ecological movies, not relying on the hypermedia model, becomes possible.

4.5 Multidimensional Simulation

Multidimensional simulation was proposed by Camara *et al.* (1990) to model water resources problems. The central idea behind it is that models should include numerical, linguistic, and pictorial entities and operations whenever appropriate.

The assumption is that abstract concepts can only be represented by words and spatial phenomena are better illustrated pictorially. Linguistic and pictorial formulations should also be manipulated with linguistic and pictorial operations, at least at the user interface level. These formulations can be more intuitive than numerical manipulations and do not require the estimation of mathematical functions and parameters. The approach was also based on the principle that numerical, pictorial, and linguistic representations of reality complement each other.

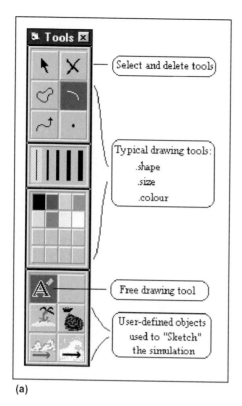

(a)

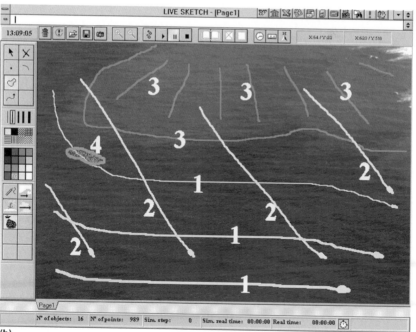

(b)

Fig. 4.19 Sketching in order to trigger an oil spill model: (a) sketching palette; (b) wind and currents (1 and 2); an island (3); and an oil spill (4) are drawn on top of a satellite; and (c) oil spill evolving over time

(c)

Fig. 4.19 *Continued*

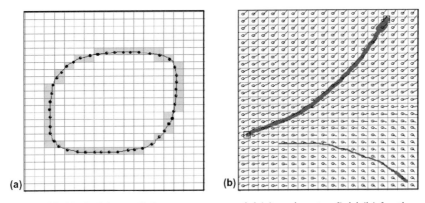

(a)   (b)

Fig. 4.20 Underlying cellular automata model (a) and vector field (b) for the oil spill model

Traditionally, linguistic variables are translated into numerical intervals or fuzzy numerical sets (Wenstop, 1976). Pictorial entities are decoded into numerical representations, which are often manipulated using numerical functions. The proposed approach (called IDEAS, after Integrated DEcision Aiding

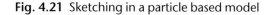

Fig. 4.21 Sketching in a particle based model

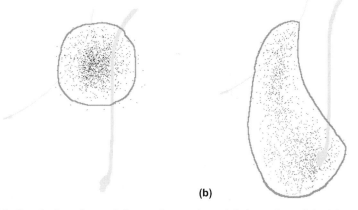

(a)   (b)

Fig. 4.22 Overlaying shape deformation and particle based models: (a) start
of the simulation; (b) after 40 simulation periods

Simulator) did not decode linguistic and pictorial entities into numerical quantities. However, it applied some conventions regarding linguistic and pictorial
entities:

- Linguistic variables have linguistic values, which may be categorical or
  ordinal. Categorical values reflect conventions such as the soil type. Ordinal
  values are defined by associating consensual interval scales to linguistic
  values (i.e., the use of pH interval scales).

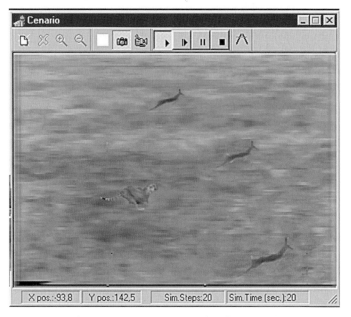

Fig. 4.23 Video output of a PbR model

- Pictorial entities could be of three types, following Arnheim (1969): pictographs (representing real objects), symbols (abstract representations of objects or processes), and signs. These pictorial entities may be characterised by four variables: size, position, shape, and colour.

IDEAS included numerical, linguistic, and pictorial modelling modes as well as the possibility of modes where numerical, linguistic, and pictorial entities could interact. The numerical modelling mode was based on differential equations. The linguistic mode used dynamic implication statements defining the state in time step t + dt, as a function of the linguistic state in L(t), the input or action U(t), and the current time step t:

$$F(L(t), U(t), t) => L(t + dt)$$

An example of a dynamic statement can be: if the soil pH is neutral at t and there is any acidic rain, then the pH of the soil at t + dt becomes acidic.

Dynamic statements are defined by experts as in a common expert system. They include action and memory components. The action component is similar to the rate terms in system dynamics models. Experts also establish transition rules as in Fig. 4.24 to define this component.

The memory term represents the accumulation process present in finite difference calculus. To represent this term, Camara *et al.* (1990) used a threshold

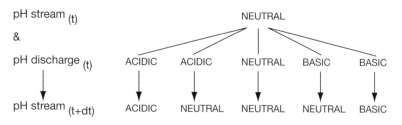

Fig. 4.24 Rules to determine the action contribution in a linguistic model

Source: Camara *et al.* 1990; reprinted with permission

rate (i.e., the number of previous integration steps) which are relevant to determine current variable values. The pictorial modelling used the concepts explained in the programming by reproduction stage (see Section 4.4). It can also apply approaches inspired on individual based animation models. In general terms, each pictorial entity was considered as being a living organism defined by a string of properties (colour, size, shape, position). A pictorial entity may reproduce itself, mutate (change of properties), or encounter other entities. The encounters could be fertile (producing offspring that could inherit properties of their parents) or sterile (i.e., an entity could absorb other entity). The iconic notation for a pictorial calculus is shown in Table 4.3.

To model the interactions between numerical, linguistic, and pictorial entities, transition rules may depend on the nature of the linguistic entities (categorical or ordinal). Figures 4.25 and 4.26 illustrate concepts proposed by Camara *et al.* (1990). Figure 4.27 shows the screen output of an IDEAS model for an oil spill. Its underlying causal diagram is shown in Fig. 4.28.

IDEAS is a multidimensional simulation engine. The use of linguistic, pictorial and other multidimensional entities (such as sound) also benefit environmental simulations in other stages of the modelling process, namely, in the calibration, validation, analysis of results, representation of uncertainty, and user interface design. These topics are covered in the next two chapters.

4.6 Further Exploration

The book site (http://gasa.dcea.fct.unl.pt/camara/index.html) will update the sites listed below. It also includes student projects on environmental modelling.

Environmental Modelling: Generic Sites

For a review of ecological models see the Register of Ecological Models (http://dino.wiz.uni-kassel.de/ecobas.html).

Table 4.3 Pictorial simulation rules

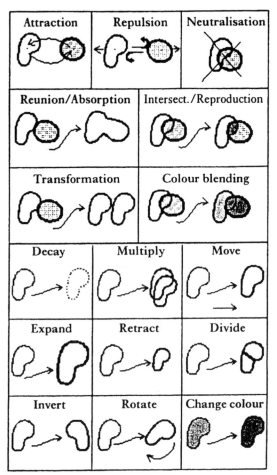

A comprehensive list of commercial software for water, air, and soil quality modelling and solid waste planning is provided at Scientific Software (http://www.scisoftware.com) and at Boss International (http://www.bossintl.com).

EPA software is available at (http://www.epa.gov/epahome/dmedia.htm).

For a library of Excel and Visual Basic models for environmental engineering, planning and management developed by students at the New University of Lisbon (http://gasa3.dcea.fct.unl.pt/assa).

For desktop tools for statistical, simulation, optimisation, and decision modelling see the Palisade catalogue (http://www.palisade.com).

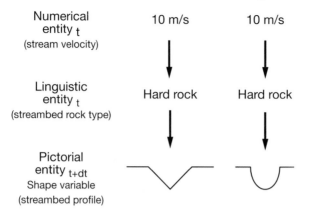

Fig. 4.25 Interaction between numerical, categorical linguistic, and pictorial entities

Source: Camara *et al.* 1990; reprinted with permission

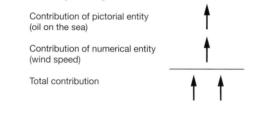

Fig. 4.26 Interaction between ordinal linguistic, pictorial, and numerical entities

Source: Camara *et al.* 1990; reprinted with permission

For interactive environments providing a programming language, built-in primitives, and graphical capabilities, see:

- Maple (http://www.maplesoft.com)
- Mathematica (http://www.wri.com)
- Matlab (http://www.mathworks.com)
- Mathcad (http://www.mathsoft.com)
- Excel (http://www.microsoft.com)

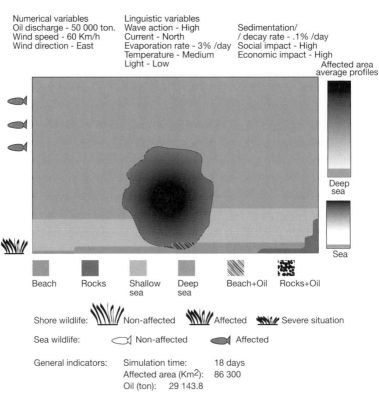

Numerical variables
Oil discharge - 50 000 ton.
Wind speed - 60 Km/h
Wind direction - East

Linguistic variables
Wave action - High
Current - North
Evaporation rate - 3% /day
Temperature - Medium
Light - Low

Sedimentation/
/ decay rate - .1% /day
Social impact - High
Economic impact - High

Affected area
average profiles

Deep
sea

Sea

Beach   Rocks   Shallow   Deep   Beach+Oil   Rocks+Oil
                        sea   sea

Shore wildlife: Non-affected   Affected   Severe situation

Sea wildlife: Non-affected   Affected

General indicators: Simulation time: 18 days
                      Affected area (Km^2):   86 300
                      Oil (ton):   29 143.8

Fig. 4.27 IDEAS model for an oil spill

Source: Camara *et al.* 1990; reprinted with permission

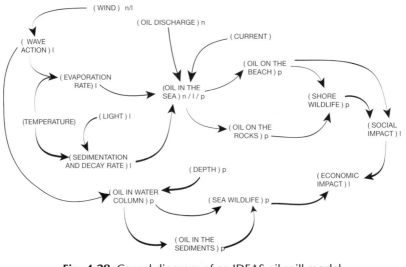

Fig. 4.28 Causal diagram of an IDEAS oil spill model

Source: Camara *et al.* 1990; reprinted with permission

For software on operations research and mathematical modelling (http://www.informs.org).

For subroutine libraries see Numerical Recipes (http://www.nr.com). A free library of mathematical software is provided at (http://www.netlib.org).

Water Quality Models

See the USGS site (http://www.gwsoftware.com/usgs.htm) and the EPA site mentioned above.

Commercial software is presented in two European sites:

- Wallingford (http://www.nerc-wallingford.ac.uk/ih/)
- Danish Hydraulic Institute (http://www.dhi.dk)

Air Quality Models

Visit the EPA site (see above), but also the Community Modelling and Analysis System, which includes an environmental decision support (http://www.iceis.mcn.c.org/CMAS).

Soil Pollution Models

Commercial software is available at Envirosoft (http://www.mmenvirosoft.com).

Noise Pollution Models

The Institute for Noise Control Engineering (http://www.ince.org) provides access to information on noise models.

Climate Models

Excellent lectures on climate models are available (http://www.geog.ox.ac.uk/students/lectures).

Visualisation and Optimisation

Chris Jones' Visualization and Optimization Web site is at (http://www.chesapeake2.com/itorms).

Cellular Automata

Santa Fe Institute maintains an excellent site (http://alife.santafe.edu/alife/topics/ca).

For a downloadable application of cellular automata to land use planning developed by Keith Clarke and his team (http://www.ncgia.ucsb.edu/projects/gig).

Individual Based Modelling

A review site of individual based models was developed by Craig Reynolds and is available at (http://hmt.com/cwr/ibm.html).

The SWARM project is located (http://www.swarm.org). SWARM provides free tools for multi-agent simulation.

To read about the Gecko simulator see (http://peaplant.biology.yale. edu:8001/papers/swarmgecko/rewrite.html).

A description and a gallery of images on the Virtual Fish Tank is available at (http://el.www.media.mit.edu/groups/el/projects/fishtank/).

To read about the Ecosim system see (http://www.isima.fr/ecosim).

Biota.org maintains a site with a retrospective of Karl Sims' work with particle systems and evolving creatures (http://www.biota.org/conf97/ksims.html). At this site also visit the Nerve Garden (http://www.biota.org/nervegarden).

Two other worthy artificial life related sites are Tom Ray's (http://www.hip.atr.co.jp/~ray) and Jeffrey Ventrella's (http://www.ventrella. com). The latter site enables the download of free artificial life software.

Programming by Demonstration and Reproduction

To learn more about programming by demonstration visit the site associated with Cypher's book (http://dnai.com/~cypher/WatchWhatIDo) and Henry Lieberman's site (http://lieber.www.media.mit.edu/people/lieber).

The work on sketching, Programming by Reproduction, and interactive ecological movies done by Edmundo Nobre is available at (http://gasa. dcea.fct.unl.pt/gasa/gasa98.htm).

References

Albini, F.A. (1976). *Estimating Wildlife Behavior and Effects.* USDA Forest Service, General Technical Report INT-30.

Arnheim, R. (1969). *Visual Thinking.* Berkeley: University of California Press.

Beck, H. and Fishwick, P. (1989). Incorporating Natural Language Descriptions into Modelling and Simulation. *Simulation,* 53/3: 102–9.

Bossel, H. (1994). *Modelling and Simulation.* Wellesley, MA: A.K. Peters.

Brebbia, C., Kenny, J., and Ciskowski, C.D. (1997). *Computational Accoustics and its Environmental Applications.* Southampton, UK: Computational Mechanics Publications.

Brown, D. and Rothery, P. (1993). *Models in Biology.* New York: Wiley.

Brown, W.M., Mitasova, H., and Mitas, L. (1997). Design, Development and Enhancement of Dynamic Multidimensional Tools in a Fully Integrated Fashion with the GRASS GIS. *Report for USA CERL.* University of Illinois, Urbana-Champaign, IL (http://www2.gis.uiuc.edu/erosion/gsoils/vizrep2.html).

Bryson, S. and Levit, C. (1992). Virtual Wind Tunnel: an Environment for the Exploration of Three-Dimensional Unsteady Flows. *IEEE Computer Graphics and Applications*, 12/4: 25–34.

Burrough, P.A. (1997). Environmental Modelling with Geographical Information Systems. In Z. Kemp (ed.), *Innovations in GIS 4*, 143–53. London: Taylor & Francis.

Buxton, W. (1989). Introduction to this special issue on nonspeech audio. *Human Computer Interaction*, 4/1: 1–9.

Camara, A.S., Ferreira, F.C., and Castro, P. (1996). Spatial Simulation Modelling. In H. Scholten, M. Fisher, and D. Unwin (eds), *Spatial Analytical Perspectives in Social and Environmental Sciences*, 201–12. London: Taylor & Francis.

Camara, A.S., Ferreira, F.C., Diogo, P., Gonçalves, P., and Silva, J.P. (1993). Multimedia System Dynamics Models for Environmental Education. In E. Kerchoffs (ed.), *Proceedings of the European Simulation Symposium*, 277–80. Delft: Society for Computer Simulation.

Camara, A.S., Ferreira, F.C., Loucks, D.P., and Seixas, M.J. (1990). Multidimensional Simulation Applied to Water Resources Management. *Water Resources Research*, 26/9: 1877–86.

Camara, A.S., Ferreira, F.C., Nobre, E., and Fialho, J.E. (1994). Pictorial Modelling of Dynamic Systems. *System Dynamics Review*, 10/4: 361–8.

Camara, A.S., Neves, J.N., Muchaxo, J., Fernandes, J.P., Sousa, I., Nobre, E., *et al.* (1998). Virtual Environments and Water Quality Management. *Journal of Infrastructure Systems, ASCE*, 4/1: 28–36.

Camara, A.S., Pinheiro, M., Antunes, M.P., and Seixas, M.J. (1987). A New Method for Qualitative Simulation of Water Resource Systems: 1. Theory. *Water Resources Research*, 23/11: 2015–18.

Campos, A.M. and Hill, D.R.C. (1999). An Agent-based Framework for Visual-Interactive Ecosystem Simulation. *Transactions of the Society for Computer Simulation*, 15/4: 139–52.

Canfield-Smith, D., Cypher, A., and Spoher, J. (1994). Programming Agents without a Programming Language. *Communications of the ACM*, 37/7: 55–67.

Castro, P. (1996). Dynamic Water Quality Modelling Using Cellular Automata. PhD Thesis, Virginia Tech, Blacksburg, VA.

Chapra, S. (1997). *Surface Water Quality Modelling*. New York: McGraw-Hill.

Chen, J.X., Lobo, N.D., Hughes, C.E., and Moshell, J.M. (1997). Real-Time Fluid Simulation in a Dynamic Virtual Environment. *IEEE Computer Graphics and Applications*, 17/3: 52–61.

Clark, M. (1996). *Transport Modelling for Environmental Engineers and Scientists*. New York: Wiley.

Coyle, R.G. (1983). The Technical Elements of the System Dynamics Approach. *European Journal of Operational Research*, 14/4: 359–79.

Cypher, A. (ed.) (1993). *Watch What I Do: Programming by Demonstration*. Cambridge, MA: MIT Press.

Damer, B., Gold, S., Marcelo, K., and Revi, F. (1999). Inhabited Virtual Worlds in Cyberspace. In J.C. Heudin (ed.), *Virtual Worlds*, 127–52. Reading, MA: Perseus.

DeAngelis, D.L. and Gross, L.J. (eds) (1992). *Individual-Based Models and Approaches in Ecology*. London: Chapman & Hall.

Dusenbery, D. (1992). *Sensory Ecology*. New York: W.H. Freeman.

Fedra, K. and Loucks, D.P. (1985). Interactive Computer Technology for Planning and Policy Modelling. *Water Resources Research*, 21/2: 114–22.

Fedra, K. (1997). Integrated Environmental Information Systems: From Data to Information. In N.B. Harmancioglu, M.N. Alpaslan, S.D. Ozkuç, and V.P. Singh (eds), *Integrated Approach to Environmental Data Management Systems*, 367–78. Dordrecht: Kluwer.

Flake, G.W. (1998). *The Computational Beauty of Nature*. Cambridge, MA: MIT Press.

Friedrich, H., Munch, S., and Dillman, R. (1996). Robot Programming by Demonstration (RPD): Supporting the Induction by Human Interaction. *Machine Learning*, 23/2–3: 163–89.

Gaither, K., Moorhead, R., Nations, S., and Fox, D. (1997). Visualizing Ocean Circulation Models Through Virtual Environments. *IEEE Computer Graphics and Applications*, 17/1: 16–19.

Garrote, L. and Bras, R.L. (1993). Real-Time Modelling of River Basin Response Using Radar-Generated Rainfall Maps and a Distributed Hydrologic Database. *Report No. 337*. Cambridge, MA: MIT.

Gershenfeld, N. (1999). *The Nature of Mathematical Modelling*. Cambridge, UK: Cambridge University Press.

Gomes, J.M. (1997). Artificial Fish in a Virtual Environment. Unpublished internal report, Department of Science, New University of Lisbon, Monte de Caparica, Portugal.

Gonçalves, P. (1997). Structure and Scale of Forest Fire Patterns. *Seminario de Revisao de Projectos de Doutoramento, Grupo de Análise de Sistemas Ambientais, FCT/UNL*.

Gonçalves, P. and Antunes, M.P. (1995). Multiscale Dynamic Ecological Modelling. *Proceedings of the 1st Conference on Spatial Multimedia and Virtual Reality*, Lisbon, Portugal.

Gonçalves, P. and Diogo, P. (1994). Geographic Information Systems and Cellular Automata: a New Approach to Forest Fire Simulation. *Proceedings of the European Conference on Geographical Information Systems (EGIS 94)*, 702–12, Paris.

Gonçalves, P., Neves, J.N., Silva, J.P., Muchaxo, J., and Camara, A. (1997). *Interoperability of Geographic Information: From the Spreadsheet to Virtual Environments*. International Conference and Workshop on Inter-operating Geographical Information Systems, NCGIA, Santa Barbara, CA.

Goodchild, M.F., Steyaert, T., and Parks, B.O. (eds) (1993). *Geographic Information Systems and Environmental Modelling*. New York: Oxford University Press.

Granieri, J. and Badler, N.I. (1994). Simulating Humans in VR. In R. Earnshaw, H. Jones, and J. Vince (eds), *Virtual Reality and its Applications*. New York: Academic Press.

Grossman, W.D. and Eberhardt, S. (1992). Geographical Information Systems and Dynamic Modelling. *Annals of Regional Science*, 26: 53–66.

Hogeweg, P. (1988). Cellular Automata as a Paradigm for Ecological Modelling. *Applied Mathematics and Computation*, 27: 81–100.

Itami, R.M. (1994). Simulating Spatial Dynamics: Cellular Automata Theory. *Landscape and Urban Planning*, 30: 1–2.

Jones, C. (1996). *Visualization in Optimization*. Norwell, MA: Kluwer (also http://www.chesapeake2.com/itorms).

Karfylidis, F. and Thanailakis, A. (1997). A Model for Predicting Forest Fire Spreading Using Cellular Automata. *Ecological Modelling*, 99/1: 87–97.

Laurini, R. and Thompson, D. (1992). *Fundamentals of Spatial Information Systems*. London: Academic Press.

Lieberman, H. (1994). User Interface for Knowledge Extraction from Video. *Proceedings of National Conference of the American Association for Artificial Intelligence*, 527–35.

Lieberman, H. (1995). A Demonstrational Interface for Recording Technical Procedures by Annotation of Videotaped Examples. *International Journal Human–Computer Systems*, 43/3: 383–417.

Lieberman, H. (1998). Integrating User Interface Agents with Conventional Applications. *Knowledge-Based Systems*, 11/1: 15–23.

Loucks, D.P., Taylor, M.R., and French, P.N. (1985). Interactive Data Management for Resource Planning and Analysis. *Water Resources Research*, 21/2: 131–42.

McGuffie, K. (1997). *A Climate Modelling Primer*. Chichester, UK: Wiley.

Miller, G.S.P. (1988). The Motion Dynamics of Snakes and Worms. *Computer Graphics*, 22: 169–77.

Mitas, L., Brown, W.M., and Mitasova, H. (1997). Role of Dynamic Cartography in Simulations of Landscape Processes Based on Multi-Variate Fields. *Computer & Geosciences*, 23/4: 437–46.

Mitasova, H., Mitas, L., Brown, W.M., Gerdes, D.P., and Kosinovsky, I. (1995). Modelling Spatially and Temporally Distributed Phenomena: New Methods and Tools for Grass GIS. *International Journal of GIS*, 9/4: 443–6.

Moore, I.D., Grayson, R.B., and Ladson, A.R. (1991). Digital Terrain Modelling: A Review of Hydrological, Geomorphological and Biological Applications. *Hydrological Processes*, 5: 3–30.

Nickerson, R. (1988). Counting, Counting and the Representation of Numbers. *Human Factors*, 30/2: 181–99.

Nobre, E. and Camara, A. (1999). Spatial Simulation by Sketching. In A.S. Camara and J. Raper (eds), *Spatial Multimedia and Virtual Reality*, 103–10. London: Taylor & Francis.

Nobre, E. and Camara, A. (2001). Simulation by Reproduction: The Use of Digital Video. Submitted to *Transactions of the Society for Computer Simulation*.

Peitgen, H.O., Jurgens, H., and Saupe, D. (1992). *Chaos and Fractals*. New York: Springer Verlag.

Phipps, M.J. (1992). From Local to Global: The Lesson of Cellular Automata. In D.L. DeAngelis and L.J. Gross (eds), *Individual-Based Models and Approaches in Ecology*. London: Chapman & Hall.

Prusinkiewicz, P. and Lindenmayer, A. (eds) (1990). *The Algorithmic Beauty of Plants*. New York: Springer Verlag.

Ray, T.S. (1991). An Approach to the Synthesis of Life. In C.G. Langton, C. Taylor, J.D. Farmer, and S. Rasmussen (eds), *Artificial Life: II*. Redwood City, CA: Addison-Wesley.

Resnick, M. (1994). *Turtles, Termites and Traffic Jams*. Cambridge, MA: MIT Press.

Reynolds, C.W. (1982). Computer Animation with Scripts and Actors. *Computer Graphics*, 16/3: 289–96.

Reynolds, C.W. (1987). Flocks, Herds and Schools: A Distributed Behavioral Model. *Computer Graphics*, 21/4: 25–34.

Rohr, G. (1986). Using Visual Concepts. In S.K. Chang (ed.), *Visual Languages*. New York: Plenum.

Romao, T., Dias, A.E., Camara, A.S., Buurman, J., and Scholten, H. (1998). New Tools for Coastal Managers. Presented at GIS Planet, Lisbon.

Rothermel, R. (1983). *How to Predict the Spread and Intensity of Forest Fire and Range Fires*. Rep. INT-143, US Department of Agriculture, Forest Service, Intermountain Forest and Range Experiment Station, Odgen, UT.

Schnoor, J.L. (1996). *Environmental Modelling*. New York: Wiley.

Sims, K. (1994). Evolving 3D Morphology and Behaviour by Competition. In R. Brooks and P. Maes (eds), *Artificial Life: IV*. Cambridge, MA: MIT Press.

Slothower, R.L., Schwarz, P.A., and Johnston, K.M. (1996). Some Guidelines for Implementing Spatially Explicit, Individual-Based Ecological Models within Location-based Raster GIS. In *Proceedings of the Third International Conference/Workshop on Integrating GIS and Environmental Modelling*, Santa Fé, New Mexico.

Taylor, C. and Jefferson, D. (1994). Artificial Life as a Tool for Biological Inquiry. *Artificial Life*, 1/1: 1–13.

Terzopoulos, D., Tu, X., and Grzeszczuk, R. (1994). Artificial Fishes with Autonomous Locomotion, Perception, Behaviour, and Learning in a Simulated Physical World. In R. Brooks and P. Maes (eds), *Artificial Life: IV*. Cambridge, MA: MIT Press.

Toffoli, T. and Margolus, N. (1987). *Cellular Automata Machines*. London: MIT Press.

Turner, M.G., Constanza, R., and Scalar, F. (1989). Methods to Evaluate Spatial Simulation Models. *Ecological Modelling*, 48: 1–18.

Ventrella, J. (1999). Animated Artificial Life. In J.C. Heudin (ed.), *Virtual Worlds*, 67–94. Reading, MA: Perseus.

Wenstop, F. (1976). Deductive Verbal Models of Organizations. *International Journal of Man–Machine Studies*, 8/3: 293–321.

Wheless, G. *et al.* (1996). Virtual Cheasapeake Bay: Interacting with a Coupled Physical/Biological Model. *IEEE Computer Graphics and Applications*, 16/4: 52–7.

White, R. and Engelen, G. (1997). Cellular Automata as the Basis of Integrated Dynamic Regional Modelling. *Environment and Planning B*, 24/2: 235–46.

Xie, Y.C. (1996). A Generalized Model for Cellular Urban Dynamics. *Geographic Analysis*, 28/4: 350–73.

Zannetti, P. (1990). *Air Pollution Modelling: Theories, Computational Methods and Available Software*. New York: Van Nostrand Reinhold.

5

Sensorial Exploration of Environmental Information

5.1 Foundations

Multimedia environmental information systems and models produce multi-dimensional information: symbolic and numerical data, static and dynamic images, and sounds. Scientific visualisation (and auralisation) enables the transformation of the symbolic and numerical data into geometric computer generated images (and sounds). Static and dynamic images and sounds, obtained either from the above transformation or reflecting realistic phenomena, can be explored using methods such as exploratory data analysis, data mining, animation, and real-time generation of three-dimensional graphics and sounds.

This chapter discusses the use of these techniques in the sensorial exploration of one-dimensional, and higher-dimensional environmental data. The methods reviewed rely on the creation of synthetic experiences that take into account human perceptual and cognitive capabilities, human variations, and task characteristics (Card *et al.*, 1997). These tasks include browsing and navigation, searching, comparing, grouping, analysis, and the creation of new information (Card *et al.*, 1997). More generally, sensorial exploration techniques can be used for analysis, exploration of data, decision support, and presentation purposes. Most of the techniques presented in this chapter are related to the use of vision in the analysis of information. The last part of the chapter is related to the application of hearing and other senses.

Vision is the primary sense for derivation of real-world data, and provides the bulk of knowledge about our environment. A wide range of data is assumed in most human beings to be due to previous 'visual calibration' (Wade and

Swanston, 1991). In visualisation efforts, there are three major lessons to be followed from psychology, which have been incorporated in visualisation guidelines suggested by Bertin (1981), Tufte (1983, 1990), and Marcus (1995), among others. One relates to the classic 'number seven plus or minus two', which is related to the limits on the human capacity for processing information (Miller, 1956). This principle is applied to the limited number of colours that can be used in visualisation exercises. The third set of lessons derive from the Gestalt school of psychology. Gestalt psychologists observed that the image perceived depended not only on the set of objects that constitute the image, but also on interrelationships among the objects (Goldstein, 1999). There are several relevant Gestalt principles for visualisation efforts:

- The concept of foreground and background, related to the separation that people are able to make between objects in an image (the use of desaturated colours for the background is a related recommendation).
- The grouping of objects that have similar visual characteristics (i.e., the use of colour to group similar objects).
- The grouping of objects that are closer in an image.
- The continuity principle, which means that observers tend to complete objects in an image.

Another important lesson for information representation is anecdoctal evidence that users rely on anchors they feel comfortable with. Jones (1996) mentioned that people do not have problems dealing with tables, which is evidenced by the success of spreadsheets and relational databases.

There have been many studies comparing the representation of numerical data in graphs versus tables. Lacerda (1986) tried to synthesise some of the most relevant information and conducted his own study. His most relevant conclusion was that graphs provide more holistic views of data, while tables are more adequate for specific data element retrieval.

Many of the visualisation methods referred to here go beyond providing holistic views on data. They let the user explore the data along graphical but also numerical dimensions.

5.2 Visualisation of One-dimensional Data

Shneiderman (1996), in his proposed taxonomy for information visualisation, proposes text as being one-dimensional data. Typography is the visualisation language for text.

Spiekermann and Ginger (1993) present an introduction to typography and non-rigid rules for legibility and effectiveness. In general, the number of type fonts and sizes should be limited as the use of type weights and styles (i.e., bold,

Informal011 BT

OrigGarmnd BT

EXCITEMENT

Matisse ITC

Fig. 5.1 Emotions and typefaces

italics, underlining). There should also be space between individual letters and around words and there is no point in justifying text to the right of a document (it slows reading). Spiekermann and Ginger argue that the typeface used conveys character to a document, a principle that can be readily tested in any word processor with the available fonts. Figure 5.1 illustrates this concept.

Most univariate data related to environmental phenomena are associated with data distributions (i.e., hydrological data). Histograms and bar graphs (Fig. 5.2) are traditionally used to represent distributions but there are better methods to show the data structure such as quantile plots and box plots (Cleveland, 1993). The concept of quantiles is essential for the visualisation of these functions (Cleveland, 1993): the f quantile, q(f), of a set of data is a value with the property that approximately a fraction f of the data is less than or equal to q(f). In a quantile plot, the values x(i) are graphed against f(i) as in Fig. 5.3a. This method enables the evaluation of distributions as f-values provide a basis for comparison. This can be done in quantile-quantile plots or q-q plots: graphing quantiles of one distribution against the corresponding quantiles of the other compares two distributions. Another method of comparing distributions is by using box plots (Tukey, 1977). Figure 5.3b shows that the box is a measure of the spread of a distribution. Figure 5.4 shows how distributions can be evaluated by comparing box plots.

5.3 Visualisation of Higher-dimensional Data

There are three types of higher-dimensional data relevant for environmental systems: numerical data; spatial representations, such as maps, digital terrain models, aerial photos, and satellite images; and networks.

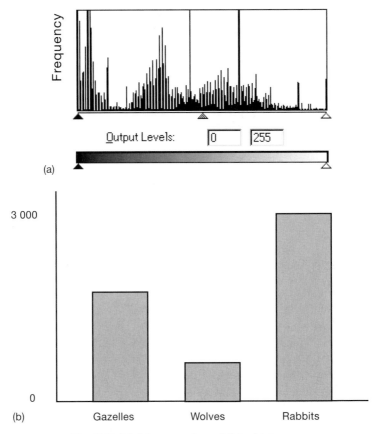

Fig. 5.2 (a) A bar graph and (b) a histogram

5.3.1 Visualisation of Numerical Data

Statistical Graphics Approaches

The most widely used tool to represent bi-variate data is the scatterplot (Fig. 5.5). Exploratory Data Analysis (EDA), first proposed by Tukey (1977), enables the brushing of scatterplots (Becker and Cleveland, 1987; Becker *et al.*, 1987), that is, users can select points on the two-dimensional plots and see their positions in other views simultaneously. In addition, EDA tools enable one to find patterns, clusters, correlations, gaps, and outliers (Shneiderman, 1996). (Web addresses for EDA and other statistical graphics tools are included in Section 5.8.)

Many problems involve three or more variables. A statistical graphics method uses point clouds (such as in scatterplots) but also surfaces to represent data in three or more dimensions. In the latter case, data can be projected into three (or

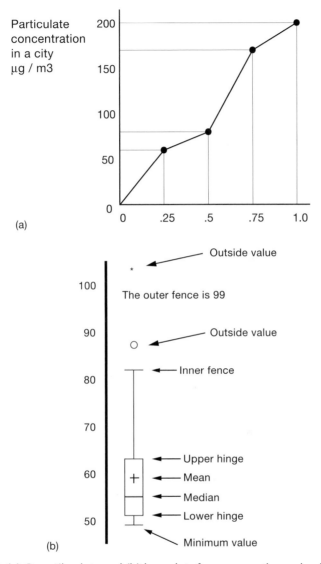

Fig. 5.3 (a) Quantile plots and (b) box plots for representing univariate data

lower) dimensions using multiple views or a dynamic sequence of views through the higher-dimensional space. Innovative schemes addressing multidimensional data include:

- Ahlberg and Shneiderman's (1994) dynamic querying system.
- Inselberg and Dimstale (1987) parallel coordinate plots.

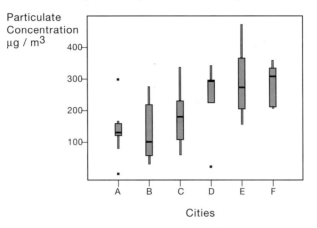

Fig. 5.4 Comparing distributions by analysing box plots

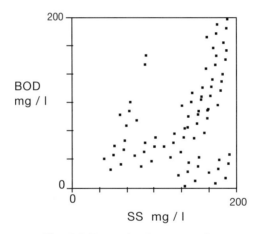

Fig. 5.5 Example of a scatterplot

- The 'worlds-within-worlds' proposed by Feiner and Beshers (1990*a*,*b*) and Beshers and Feiner (1993).
- Cook *et al.*'s (1998) use of Cave Automatic Virtual Environment (CAVE) 2 for interaction with environmental data.

Shneiderman (1996) argues that with three-dimensional scatterplots, users lose a sense of direction, and in many instances, data points are not visible as they are occluded by others. Ahlberg and Shneiderman (1994) proposed multidimensional scatterplots with each additional dimension controlled by a slider to access dynamically a spatial database. This principle is illustrated in Fig. 5.6. Spotfire

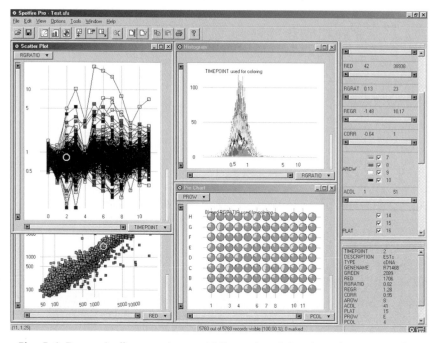

Fig. 5.6 Dynamically accessing multidimensional data by using scatterplots and sliders

Source: http://www.spotfire.com. Image courtesy of Spotfire, Inc.

software (see address in Section 5.8) is based on this concept. The work of Ahlberg and Shneiderman (1994) can be applied to solve any environmental problem where there is a set of impacts, goals, or constraints that can be modelled using sliders. Environmental impact assessment and ecological design of new products are illustrative examples of such problems.

The use of parallel coordinate plots enables the exploration of problems with a number of dimensions limited only by the size and resolution of the monitor (Inselberg and Dimsdale, 1987, 1994). This method enables the representation of the environmental impacts of alternative projects as shown in Fig. 5.7.

The use of the 'worlds-within-worlds' scheme for visualising multivariate functions relies on holding constant one or more independent variables. This means in Beshers and Feiner's (1993) words 'in taking an infinitely thin slice of the world perpendicular to the constant variable's axis, reducing the world's dimension'. By reducing the problem to three dimensions (3D), the resulting slice can be manipulated and displayed in 3D. To retrieve the higher dimensions, a 3D world is embedded in another 3D world. The position of the embedded world's origin relative to the containing world's coordinate system specifies the values of up to three variables that were held constant in the process of slicing.

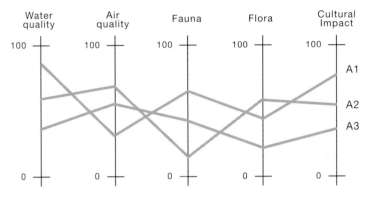

Fig. 5.7 Example of the application of parallel coordinate systems to represent environmental impacts of alternatives A$_i$

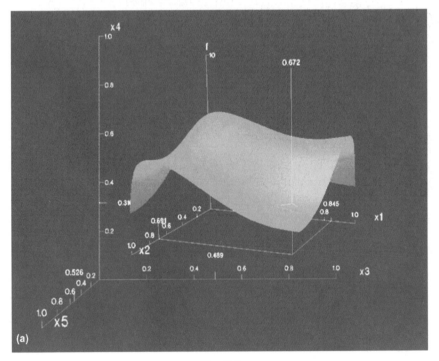

Fig. 5.8

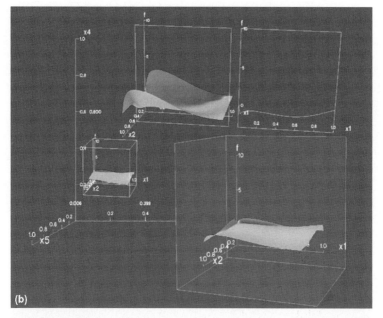

Fig. 5.8 Visualising n dimension problems using worlds-within-worlds:
(a) For a function f ($\times 1$, $\times 2$, $\times 3$, $\times 4$, $\times 5$), three variables ($\times 3$, $\times 4$, $\times 5$) are kept constant with values c3, c4, c5. This results in a new function f': f'($\times 1$, $\times 2$) = f($\times 1$, $\times 2$, c3, c4, c5). The function f' is easy to graph in 3D as a surface plot, with $\times 1$ on the *x*-axis, $\times 2$ on the *z*-axis, and the value of the function on the vertical axis (*y*-axis).
(b) Nested worlds. Only the position of the origin of the inner world relative to the outer world affects the values of c3, c4, c5, and hence the values displayed in the surface plot. Both the inner and outer worlds may be scaled and rotated for better viewing

Source: figs 1 and 2 of AutoVisual
(http://www.cs.columbia.edu/graphics/projects/AutoVisual/AutoVisual.htm). Copyright 1993,
Clifford Beshers and Steven Feiner, Columbia University

This process can then be repeated by additional recursive nesting (Beshers and Feiner, 1993). Figure 5.8 illustrates the method.

The Cook *et al.* (1998) use of CAVE 2, a highly immersive virtual reality environment (see the introduction in Section 5.6) is illustrated in Fig. 5.9. The system enables the exploration of water chemistry data from sampling sites. The viewing box shows the scatterplot of measured variables. Interaction with data in the 3D environment is shown in Fig. 5.10. In this environment, one may paintbrush data with different colours and geometric representations. Any viewpoint can be also achieved. This eliminates Shneiderman's occlusion objection to three-dimensional scatterplots.

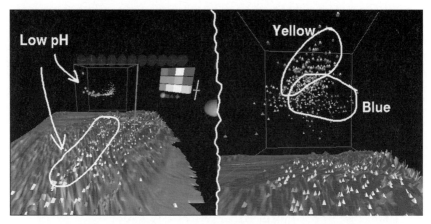

Fig. 5.9 Display of water chemistry data in CAVE 2
Source: http://www.public.iastate.edu/~dicook. Image courtesy of Dianne Cook

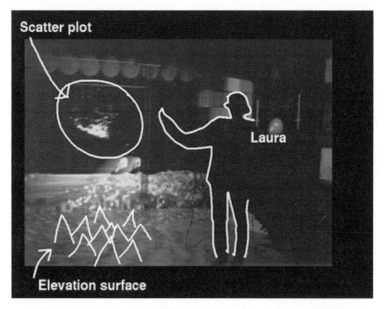

Fig. 5.10 Interaction with water chemistry data in CAVE 2
Source: http://www.public.iastate.edu/~dicook. Image courtesy of Dianne Cook

Statistical graphical methods for exploring data are also used for data mining. Data mining is an automated process for discovering information on data warehouses. Data mining includes classification, clustering, and visualisation methods (Berson and Smith, 1997; Groth, 1998). In classification methods (or supervised learning), there is always one dependent variable and the goal is to discover factors that influence that variable value. In clustering methods (or non-supervised methods), there are no dependent variables and one looks to the grouping of data that share identical patterns or trends.

Visualisation is applied in data mining to explore the database, display the results obtained with classification or clustering methods, and as a verification tool for the data mining process. Silicon Graphics Mineset (see the Web address in Section 5.8) illustrates many of the concepts associated with visual data mining. Mineset uses decision trees as their data mining engine. Decision trees may be seen as predictive models that attempt to determine factors explaining values of dependent variables (Quinlan, 1992), where:

- Data is divided by each tree branch without loss of information.
- Segmentations are carried out in each branch by defining questions relevant to the data.
- The algorithm used in decision trees selects the questions that will better subdivide data in homogeneous segments.
- CHAID (Chi Square Automatic Interaction Detector) applies the chi-square test and considers variables defined as categories.
- Decision trees stop whenever: a segment as a minimum predefined number of records; a segment is organised in a sole prediction value; or the improvement obtained by further segmentation is irrelevant.

Mineset includes other visualisation tools such as the Scatter Visualiser and Map Visualiser. The Scatter Visualiser, an extension of Ahlberg and Shneiderman's (1994) proposal (see Fig. 5.6), enables the visualisation of up to nine dimensions by using 3D coordinate systems, two independent sliders, size, colour, and orientation. By changing the values of the variables in the two sliders, the values of independent data are modified. These visualisers enable the immediate change in the size, colour, and positions of the entities representing the dependent data. Filters can also be used to display only the entities meeting certain criteria.

With the Map Visualiser, strong spatial relationships can be explored. Data is displayed as graphical elements on a map (Fig. 5.11), with user defined variables indicated by the height and colour of each element. Users can drill down for more detailed information about specific areas.

Scientific Visualisation Approaches

The production of information visualisations involves the transformation of data into visual representations. The most common approach, in scientific visualisation, is the conversion of data into graphical primitives (points, lines, polylines, and polygons).

The conversion of data into graphics primitives includes three stages: filtering, mapping, and rendering (Foley *et al.*, 1990). Filtering is the extraction of

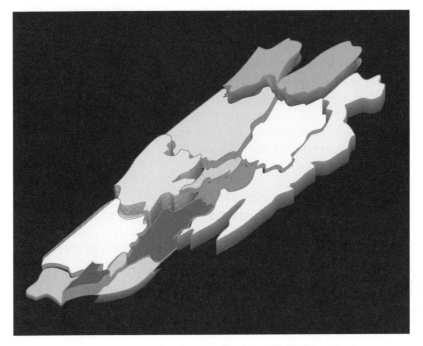

Fig. 5.11 Example of an application of Map Visualiser

features or reduction in quantity of data by computing derived quantities (i.e., derivation of a flow line from a velocity field, Rhyne, 1997). Mapping is the conversion of the resulting data into graphical primitives. Finally, by rendering one obtains the colour, illumination, shading, and textures associated with the geometric data. Visualisation of environmental data has two important properties:

- Visualisation data is discrete, because data is typically sampled at a finite number of points and computers used to process it are digital. This means that the values of regions between data points are not known. As a result, interpolation methods have to be applied, which means looking to relationships with neighbouring points using linear, quadratic, and other interpolation functions such as splines (see Appendix 3).
- Visualisation data may be regular or irregular. In the former, there is a relationship between data points such as in a structured grid. Irregular data can represent, however, information more densely when changes are intense and less densely when changes are not so severe.

Visualisation datasets have two properties (Schroeder *et al.*, 1998): structure and data attributes. The structure is characterised by topology and geometry. Topology is the set of properties that does not change with transformations such

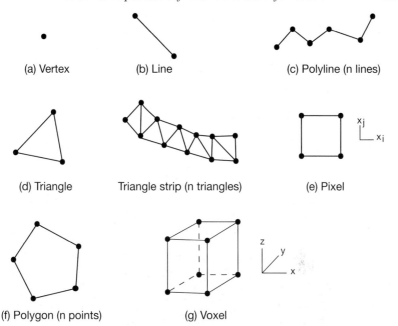

Fig. 5.12 Examples of cell types used in environmental visualisation: (a) vertex; (b) line; (c) polyline; (d) triangle; (e) pixel; (f) polygon; (g) voxel

as rotation, translation, and scaling. Geometry is the instantiation of topology (i.e., the coordinates of a polygon).

The structure of a dataset consists of cells and points (where data values are known). The cells specify the topology, while the points specify the geometry. The attributes may be associated to cells or points. Figure 5.12 shows cells commonly used in environmental visualisation. Note that topology is implicitly defined by the ordering of the point list.

Attribute data types include (Schroeder *et al.*, 1998):

- *Scalars*. Examples of scalar data are temperature and elevation, valued at points of the dataset.
- *Vectors*. Magnitude and direction define vector data. Examples are sea currents and particle trajectories.
- *Normals*. These are vectors with magnitude equal to 1. They are often used to control the shading of objects and may also be applied to control the orientation and generation of cell primitives (see Appendix 3).
- *Texture*. This is defined by regular arrays of colour, intensity, and/or transparency values that provide extra detail to rendered objects. The draping of polygons with photo textures (see draping of a digital terrain model in Fig. 5.18) is an example of texture mapping.

- *Tensors.* Tables describe tensors with dimensions specified by their rank. A tensor of rank 0 is a scalar, rank 1 is a vector, rank 2 is a matrix, and rank 3 is a 3D rectangular array. Tensors are used to represent electromagnetic fields (Santos, 1994).

The datasets used in visualisation may be classified according to their structure: regular or irregular. Regular or structured datasets can be implicitly represented in computerised visualisation systems. Irregular data must be explicitly described due to their lack of pattern. In environmental visualisation, the most common dataset types are:

- Structured point datasets, where points may be lines (1D), pixels (2D), or voxels (3D). Data are moved into these grids using interpolation and extrapolation methods. Volume rendering techniques are used to create the visual display in a process called volume rendering, whenever data is 3D. Water pollution and weather datasets may be visualised using volume rendering (see visualisation of datasets included in the Environment Expo98 demos on the book's Web site.
- Polygonal datasets, where the topology and geometry is unstructured, and may be used in architectural walkthroughs.
- Structured grids used in finite difference environmental quality models.
- Unstructured grids applied in finite element analysis in water and soil models.
- Unstructured points, reflecting the irregular location of sampling points.

The algorithms most widely used to transform data include (Schroeder *et al.*, 1998):

- *Geometric transformations* that change geometry but not topology. Examples are translation, rotation, and scaling the points of a polygonal dataset.
- *Attribute transformations* may convert attributes from one form to another or create scalars from input data (i.e., elevation).
- *Combined transformations* that change the dataset structure and attributes. An example is the computation of contour surfaces.

Algorithms may also be classified according to the type of attributes they operate on. The two most significant categories are:

- *Scalar algorithms*, such as colour mapping (mapping scalar data to colours) and the generation of contour lines or surfaces. These are visual representations of a function $f(x, y)$ (when considering only two dimensions), such that the only points drawn are those satisfying $f_i(x, y) = C_i$, where C_i are contours, curves representing specific levels in the data space (Fig. 5.13). Contour plots imply the use of interpolation methods.
- *Vector algorithms* to show a wind velocity field using arrows as in Fig. 5.14.

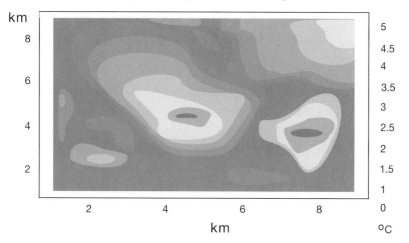

Fig. 5.13 Example of a contour plot

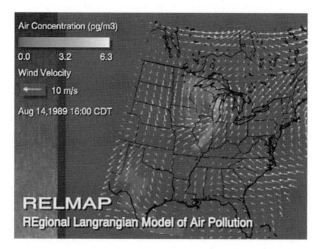

Fig. 5.14 Visualisation of vectors reflecting wind velocities
Source: http://www.epa.gov/vislab/svc/index.html

The final stage of the visualisation process is the rendering of an image. The goal of rendering here is not so much image realism or aesthetic qualities as its information content (for a comment on information content or its absence in visualisation efforts, see Globus and Raible, 1994). Rendering may produce a see-through wire frame, a hidden-line rendering (where surfaces that are occluded by other surfaces are hidden), and a shaded rendering. The latter takes into account lighting and shading and is discussed here as simply rendering.

Rendering is the result of the interplay between (see also Appendix 3):

- the three dimensional shape of the object being rendered (geometry);
- the point of view (camera);
- the lighting of the scene (lights);
- the surface characteristics of the object; and
- the rendering algorithm that will display in two dimensions the 3D object.

Most rendering algorithms are based on an approach called scan-line rendering. In this approach, the program looks at each pixel, one after the other, each row of pixels (called scan line) after the other, and calculates the colour that pixel should be rendered. To calculate the colour that a given pixel should assume, a process called ray casting can be used. From the point of view of the camera or 'eye', a ray is cast through the first pixel of the first scan line. The eye then follows the ray until the ray either hits the surface of an object or leaves the viewable scene. If the ray hits the object, the program calculates the colour of the object at the point where it has been hit. After defining the colour for the first pixel, the algorithm then moves on to the second pixel and performs the same steps. This process is repeated for all the pixels of the image. To calculate the colour for each pixel, shading algorithms are used.

Ray casting has the limitation of dealing with only one object at a time, not considering the presence of other objects in the scene. To address this problem, ray tracing is an alternative. Ray tracing deals with all the objects simultaneously, looking to the bounces of the ray. Ray tracing does not handle well the diffuse reflection of light from one surface to another. Radiosity algorithms may be then applied if realism is a goal in the visualisation effort. These algorithms work by breaking down every surface into a small number of smaller surfaces, to calculate the diffuse reflections among these surfaces (O'Rourke, 1998).

In this discussion, it is assumed that the goal is to render a surface (surface rendering). In environmental phenomena there are, however, objects, such as clouds and water, that are translucent or scatter the light that passes through them. These objects require volume rendering (i.e., the interior of the object is also to be rendered).

For the visualisation of multidimensional environmental problems, glyphs are tools often used. Their size, shape, colour, and texture can each be utilised to represent a variable in the data. Figure 5.15 illustrates the application of glyphs. Glyphs are affected by input data and they may alter the pictorial object in response to data. They may be used to represent a local distribution of values (local icons) or the structure of a complete dataset (global icons). Glyphs may be displayed as arrows, spheres, needles, or any other suitable iconic representation.

There are numerous examples of scientific visualisation of environmental phenomena on the Web sites listed in Section 5.8. Other applications have been proposed by, among others:

Fig. 5.15 The use of glyphs and slices in the visualisation of the impacts of a storm in the concentration of sediments in Lake Erie. Winds from storms produce waves, which in turn create bottom currents and turbulence that can cause contaminated bottom sediments to be re-suspended in the water column. The shallower the water, the stronger the effects of the wave action. As a result, sediment concentrations tend to be also largest in shallow sections

Source: http://www.epa.gov/vislab/svc/index.html

- Kazafumi *et al.* (1989): environmental impact assessment visualisations.
- Kruse *et al.* (1992): space imaging.
- DeGloria (1993): soil behaviour visualisation.
- Wolff and Yeager (1993): provide a review of visualisations of natural phenomena.
- Fedra (1994): with an integrated software for water and air pollution visualisation.
- Fuchs (1994): marine data visualisation.
- Koussoulakou (1994): air pollution visualisation.
- Delmarcelle and Hesselink (1995): flow visualisation.
- Owen *et al.* (1996): visualisation of groundwater systems.
- Liddel and Hansen (1997): soil ecosystem visualisation.

Mathematical modelling software, such as MATLAB and Mathematica (see the Web addresses in Chapter 4), include visualisation capabilities, such as contour plots, line trajectories, and volume slice plots (as in Fig. 5.15), with different lighting and colour models (see Appendix 3).

Most scientific visualisation software (see Web addresses in Section 5.8) include these capabilities, a variety of glyphs, vector field representations,

Fig. 5.16 Visualisation of the May–Leonard dynamical system: competition of three competing species. This system may be visualised interactively at a CAVE installation

Courtesy of Alex Bourd, George Francis NCSA

volume visualisations, and other 3D representations. The turning of lines into tubes or ribbons is one of these representations that have been used to display the evolution of populations (Fig. 5.16).

Developments on Web technology, such as VRML and Java 3D, can also be used for 3D visualisations of numerical data as discussed in Section 5.6. For a discussion on Web visualisation methods see also Rohrer and Swing (1997).

There is ongoing research on adding intelligence into visualisation software. The goal is to have the software create automatically appropriate visualisations for predefined problems and assist the user in selecting the most suitable tools, such as contour plots, isosurfaces, volume renderings, and vector representations (Baker, 1994; Rogowitz and Treinish, 1994).

5.3.2 Visualisation of Spatial Information

Maps provide the geographical data that can characterise objects on (Mac-Eachren 1995):

- their position with respect to a known coordinate system;
- their physical attributes associated with the geographical position;
- their spatial relationships with surrounding geographical features.

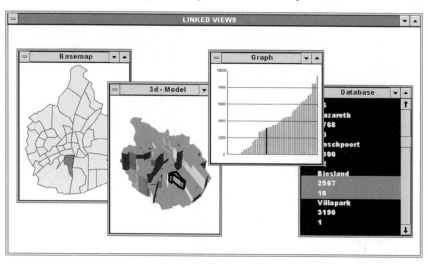

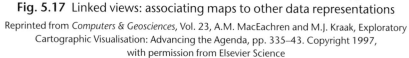

Fig. 5.17 Linked views: associating maps to other data representations

Reprinted from *Computers & Geosciences*, Vol. 23, A.M. MacEachren and M.J. Kraak, Exploratory Cartographic Visualisation: Advancing the Agenda, pp. 335–43. Copyright 1997, with permission from Elsevier Science

MacEachren (1995) and Kraak and Ormeling (1996) provide a review on traditional cartographic representations such as chloropleths, isopleths (that use the contour plot concept), dot maps and flow maps.

Aerial photos and satellite images may provide realistic visualisations and, after classification, of spectral data associated to terrains, as discussed in Chapter 2. As MacEachren and Kraak (1997) have commented, there are several trends in spatial visualisation and interaction that go beyond the use of traditional maps and remote sensing images. They include:

- The association of linked views including 3D models, graphs and databases to maps such as in Fig. 5.17. Other examples are provided by Cook *et al.* (1997) and Anselin (1999) linking mapping and exploratory data analysis software, and Shiffer (1993), augmenting geographical information with multimedia.
- The superimposition of air pollution plumes on maps, aerial photographs, or satellite images, as discussed in Boice (1992) and Chakraborty and Armstrong (1996). Monmonier (1999) presents related visualisation examples from weather forecasting (see the Website listed in Section 5.8). The book's Web site includes an illustrative QuickTime movie of this technique.
- The use of animation in dynamic mapping (see Fig. 5.22), as proposed by DiBiase *et al.* (1992) and Mitas *et al.* (1997).

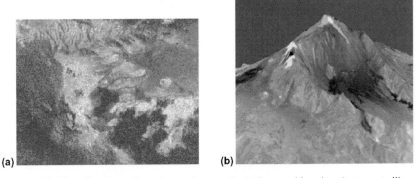

Fig. 5.18 Visualisation of a volcano is greatly enhanced by draping a satellite image (a) on to a digital terrain model (b)

Source: http://gasa.dcea.fct.unl.pt/geo@news/index.html. The image of Etna volcano is published in Realmuto *et al.*, 1994, and is reprinted with permission

- The visualisation of uncertainty of spatial information (see Section 5.6).
- The exploration of 3D representations of the terrain. These may be digital terrain models draped with photo textures as in Fig. 5.18 and in virtual reality representations as discussed in Section 5.6.

5.3.3 Visualisation of Networks

Networks are relevant for environmental applications as they represent physical phenomena (utility networks, solid waste removal routes, road networks, and river networks) and provide metaphors for non-physical data (relationships between entities in environmental information systems or models such as shown in Fig. 4.28). Networks are straightforward, easy to explain, and can be drawn and revised quickly. Network types that may be of interest include grids, trees, circuits, and weighted graphs as shown in Fig. 5.19.

However, there are some problems associated with the visualisation of networks, as pointed out by Shneiderman (1996) and Card *et al.* (1997) such as display clutter, node positioning, and the perceptual tensions occurring when nodes that are close are not related. There are several alternatives for avoiding display clutter, including:

- The associated use of adjacency or link matrices, that allow the retrieval of each individual relationship (Fig. 5.20).
- The use of lenses as proposed by Furnas (1986) and applied by Bodin and Levy (1994). Moving Filters, a technology discussed in Chapter 6, is based on this principle.

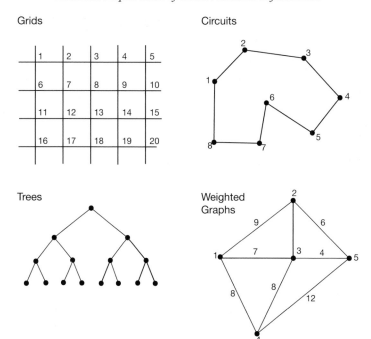

Fig. 5.19 Network types (clockwise from top left): grids (i.e., relevant for domestic solid waste collection routing, a Chinese postman problem); circuits (i.e., hospital solid waste collection, a travelling salesman problem); trees (i.e., for raster data structures); weighted graphs (i.e., shortest path problems)

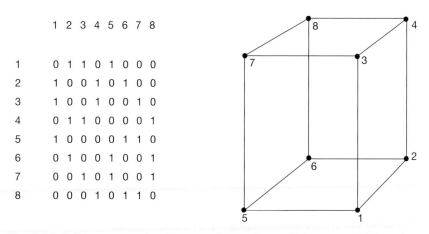

Fig. 5.20 Adjacency matrix of a network. Matrix coefficients are 1 if there is an arc connecting nodes i to j; 0 otherwise

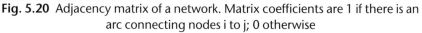

There are several additional interactive techniques for displaying a network as discussed by Jones (1996). They include:

- moving nodes or edges;
- geometric zooms or pans;
- hiding nodes or edges;
- the use of hierarchical graphs (hiding complexity, showing groupings);
- provide multiple views;
- offer database style queries besides pointing;
- use animation techniques to illustrate dynamic phenomena in a network (see Section 5.4).

5.4 Animation Methods

In the visualisation of environmental data, one is interested in observing the dynamics of:

- Numerical data visualised using the methods described above.
- Spatial phenomena represented in maps, aerial photos, or satellite images.
- Natural and artificial entities such as those represented in individual based modelling approaches.

Animation is the technique used to obtain the desired dynamic images. Animation methods of interest include keyframe animation; animation of three-dimensional entities using inverse kinematics; particle systems; and procedural animation.

5.4.1 Keyframe Animation

Traditionally, animation involves the production of a series of still images that, when played back in quick succession (usually on film or video) appear as continuously moving. This illusion is the result of a physiological feature of human vision called persistence of vision (Jones, 1996). In the early days of animation (Thomas and Johnston, 1984), the master animators designed 'key frames' of an animation sequence. The less experienced animators designed the in-betweens. Computer graphics methods still use keyframes designed by animators, but they generate the in-betweens by using linear interpolations and splines (see discussion below). In most animation efforts today, the original Disney techniques are still used such as (Jones, 1996; Laybourne, 1998; Wagstaff, 1999):

- *Squash and stretch*: objects change shape when moving.
- *Timing*: spacing actions to define the weight and size of objects and the personality of characters.

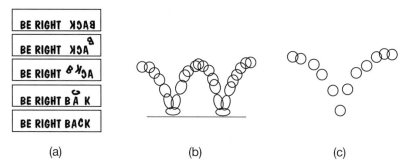

Fig. 5.21 Animation techniques: (a) anticipation; (b) squash and stretch; and (c) Slow In and Slow Out

- *Staging*: presenting an idea so that it is unmistakably clear.
- *Anticipation*: an action that makes one anticipate an event.
- *Slow In and Slow Out*: acceleration to a constant velocity and then deceleration.
- *Follow Through* and *Overlapping Action*, as not everything stops or starts moving simultaneously;
- *Arcs*, as natural objects move in arcs not straight lines.

Figure 5.21 shows examples for some of these techniques.

Keyframe animation has had numerous applications in environmental systems that may be seen in environmental visualisation sites included in Section 5.8 and on the book's Web site. Figure 5.22 illustrates the animation of maps.

An innovative use of keyframe animation was 'The Breathing Earth' shown at the Internet 1996 World Exposition (Malamud, 1997). The earthquake activity around the earth over a two-week period was displayed by spinning the world and seeing it swelling in seismic areas. Figure 5.23 shows some of the animation frames.

5.4.2 Animation of Natural and Artificial Entities

The representation of the dynamics of natural and artificial entities involves three-dimensional modelling, rendering, and setting and editing the animation.

For the modelling of entities, a surface modelling approach may be used, where the surfaces that enclose an object define the shape of the object. If needed, a solid modelling approach may also be utilised, where an object is defined as a solid mass (O'Rourke, 1998). In this case, the density, weight, and other attributes of the solid model have to be defined. For most environmental models, a surface modelling method is sufficient.

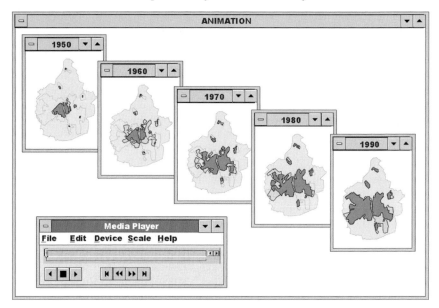

Fig. 5.22 Animation of maps

Reprinted from *Computers & Geosciences*, Vol. 23, A.M. MacEachren and M.J. Kraak,
Exploratory Cartographic Visualisation: Advancing the Agenda, pp. 335–43. Copyright 1997,
with permission from Elsevier Science

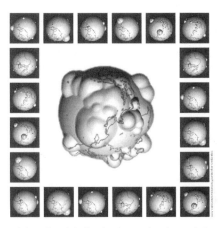

Fig. 5.23 'The Breathing Earth' displaying seismic activity around the earth.
Japan's Sensorium team developed this project for the Internet 1996 World
Exposition

Reprinted with permission from Carl Malamud

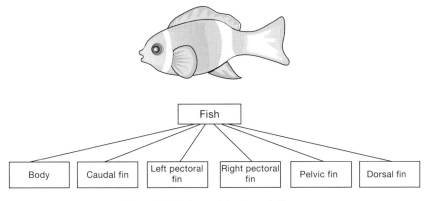

Fig. 5.24 Hierarchical modelling

Three-dimensional surface modelling is based on approximations, where a large number of small polygons approximate the curvature of the surface.

Polygonal approximations can never represent the natural curves accurately. Curves can be approximated with polylines, but this technique makes curve editing a difficult task. Most computer modelling systems now use splines (see Appendix 3).

Many modelling packages (see Section 5.8) already include geometric shapes in their libraries. They include geometric primitives such as the sphere, cube, cylinder, and other geometric objects that can be easily parameterised. They also use techniques, such as extrusion, where a curve is moved through space leaving a trace. This trace then becomes a surface. By applying transformations (such as the classic translations, rotations, and scaling), while curves are extruded through space, one obtains swept surfaces. Bending, twisting, and tapering can also deform existing objects. For complex entities, such as animals and humans, several elements are joined to represent those entities. They are typically organised into hierarchical structures, where each element is a node (Fig. 5.24).

Other techniques for modelling natural entities include plant generators, where one manipulates parameters such as the types, age, and the size and shape of plants. Many of these systems are inspired by the original work of Lindemayer reported, for instance, in Prusinkiewicz *et al.* (1990).

Fractals have also been used to generate natural entities such as landscapes. They solve simultaneously the problem of defining the three-dimensional model and rendering it (Peitgen *et al.* 1992). The three-dimensional models can then be rendered using ray tracing. In many instances, two-dimensional pictures can be added to the surface of the three-dimensional model. There are two ways to do texture mapping: either by projecting the texture image or by stretching onto the surface.

To animate hierarchical models, one positions the model for each node and sets a keyframe, then repositions the model and sets a new keyframe. This procedure is repeated for all the nodes of the model. The animation system will then interpolate in-between frames for each node.

For complex hierarchical models, such as those representing animals, this process is too laborious. A method called inverse kinematics is preferred. When animating animals (or humans), the points that move are usually at the extremities of the hierarchical model. They are called the effectors as they impact the other nodes of the model, by applying a flow of transformations (the kinematic process). The fact that this kinematic process runs in the opposite direction of the normal calculations explains why it is called inverse kinematics.

5.4.3 Particle Systems

Particle systems use a high number of simple primitives (particles) to model complex phenomena that cannot be represented by surfaces. Each particle may be defined by its position, velocity, colour, size, and lifetime. Other properties may include emission rate, turbulence, flocking and interaction with obstacles (Sims, 1990). In a particle system, the following computations are performed (Green and Sun, 1995):

- generation of new particles at the source;
- updating the particle properties;
- removing dead particles from the system;
- drawing the particles.

Particle systems were used originally to model fire (Reeves, 1983). They can be also applied to the modelling and visualisation of air and water pollutant plumes. An example of particle systems modelling in a virtual environment is described in Camara *et al.* (1998) (see video clip on the book's Website, and Fig. 7.15).

5.4.4 Procedural Animation

Particle systems are a special form of procedural animation. In this technique, one writes a program that uses procedures to generate the primitives in a model. Behaviours are usually associated with the models. The pioneer work of Reynolds (1982) and Terzopoulos *et al.* (1994), reviewed in Chapter 4, are illustrative examples. Applications of procedural animation, extensively used in films, such as *Jurassic Park* and *The Lion King*, are discussed in Magnenat-Thalman and Thalman (1994).

To obtain realistic animation, a relevant feature is the modelling of motion dynamics. Traditional motion capture techniques, called rotoscoping, aligned the computer model frames to a sequence of real-world images (from photos or video), to improve the realism of the computer representations. Channel animation is an improved method, where movements of an input device, such as a mouse, are linked to a few animation parameters such as objects rotations and translations. Currently, motion capture involves the use of sensors that wire actors. Movements of these actors are first captured in the sensors. Then, they are channelled to control the actions of computer generated characters (O'Rourke, 1998).

Animation packages are also increasingly integrating scene files that enable the animation of natural phenomena. These files are verbal and numerical descriptions of the models, the light, the surface characteristics, the camera, and the animation that the models can have (O'Rourke, 1998).

5.4.5 Animation on the World Wide Web

Wagstaff (1999) provides a review on the available methods for including animation on Web sites. (In Section 5.8 a link to his book's Web page is included.)

There are two major types of animation offered on the Web:

- Frame-by-frame images that are pre-rendered and downloaded to the user's computer.
- Animation generated on the fly within the Web browser environment.

Frame-by-frame approaches include animated GIFs and QuickTime movies and are available for most browsers.

Macromedia products (Shockwave and Flash) provide play on the fly. They run faster but currently require plug-ins. However, the number of Web users that are Flash-ready has grown exponentially. 'Secrets at Sea', a Flash-based animation, illustrates its potential environmental applications (see the Web address in Section 5.8).

QuickTime with the Vector Animation compressor and decompressor (codec) (Wagstaff, 1999) can now work with vectors instead of bitmaps. The results are similar to the ones obtained with Flash. This codec is lossless (animation does not lose quality after compression). Animations are also scalable: they can be stretched without any loss in quality. QuickTime also supports 'sprites' that can be created with Macromedia Director and other applications. Sprites are characters, that are independent of other characters and the background. They are stored separately and can interact with other sprites. This technique is used in the production of simulation outputs as videos, and is discussed in Chapter 4.

Java has been used in several environmentally related applications but few animation tools support it and it is slow. Dynamic Hypertext Markup Language (DHTML) is still another alternative but has suffered from Microsoft and Netscape's disagreements on how it should work.

The move to broadband (via cable modems and Asymmetric Digital Subscriber Line, ADSL), streaming animation formats such as Real Media and QuickTime (Appendix 1), and vector based compression schemes promise to raise the quality of Web animations to levels similar to those common to desktops in the near future.

5.5 Visualisation of Uncertainty

The visualisation of uncertainties associated to environmental phenomena includes the analysis of numerical data (real and simulated) and spatial representations such as maps and images. In both cases, uncertainties can be associated with:

- the data acquisition process (i.e., errors in measurement);
- the frequency of occurrence (i.e., natural accidents);
- the transformations that alter the original form of the data;
- the visualisation process itself (i.e., the use of approximate radiosity algorithms in 3D visualisations, and the errors associated with interpolations).

Pang *et al.* (1997) provide an extensive review on the visualisation of uncertainty, which includes statistical variations, errors and differences, range values, and noisy or missing data. These authors consider that there are two main attributes that can be used for the classification of visualisation methods:

- The value of a datum and its associated value uncertainty. Most environmental data values can be characterised as a scalar, multivariate, or a vector variable. The multivariate variables may include different components that may be scalars or vectors.
- Visualisation extent that may be discrete or continuous, depending on whether an individual datum or a continuous range of data are considered.

For environmental systems, relevant methods for uncertainty visualisation include:

- *Scalar values.* For discrete visualisation, glyphs, such as error bars (Fig. 5.25) and box plots (Fig. 5.3b), have been applied. As shown in Fig. 5.25, glyphs encode information by their shape and colour. For the visualisation of continuous ranges of data, one may use difference and side-by-side images and blinking, often related to the randomisation of dots designed to portray the occurrence of events or objects.
- *Multivariate values.* In discrete visualisations, Chernoff faces (Chernoff, 1971) and scatterplots (Figs 5.5 and 5.6) may be used. For continuous visualisation, side-by-side and difference images can be applied.

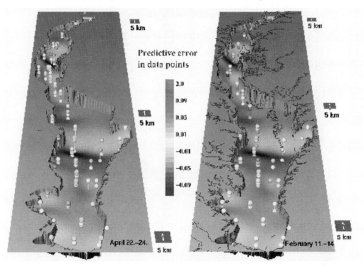

Fig. 5.25 Visualisation of errors associated with data points

Reprinted from *Computers & Geosciences*, Vol. 23, L. Mitas, W.M. Brown, and H. Mitasova,
Role of Dynamic cartography in Simulations of Landscape Processes Based on Multivariate Fields,
pp. 437–46. Copyright 1997, with permission from Elsevier Science

- *Vector values.* Glyphs may be used for the discrete case (Wittenbrink *et al.*, 1996). Modified tubes and ribbons (Fig. 5.16) can be used for continuous visualisation.

Other techniques include changes of geometry attributes, animation, and sonification, which are also suitable for the visualisation of uncertainties in spatial representations.

Fisher (1999) considers that there have been two major sources of uncertainty researched for geographical information:

- The classes of the spatial objects or the individuals are poorly defined, and there is vagueness in classifying classes or objects. This is usually managed by using fuzzy set theory.
- The spatial objects and the classes they belong to are well defined. The uncertainties are caused by errors and are probabilistic in nature.

The application of fuzzy set theory means that the degree of membership of spatial objects in certain classes is specified. Those degrees are numbers between 0 and 1. Memberships may be specified by the Similarity Relation Model or the Semantic Import Model (Robinson, 1988).

The Similarity Relation Model implies the use of data driven clustering algorithms. The Semantic Import Model relies on formulae specified by the

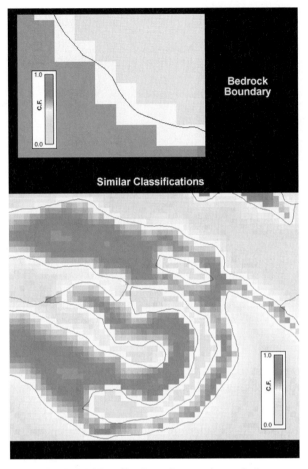

Fig. 5.26 Visualisation of vague boundaries

Reprinted from *Computers & Geosciences*, Vol. 23, T.J. Davis and C.P. Keller, Modeling and Visualizing Multiple Spatial Uncertainties, pp. 397–408. Copyright 1997, with permission from Elsevier Science

user (Fisher, 1999). Burrough (1989) and Burrough *et al.* (1992) discuss classic applications of fuzzy set theory to soil information systems. Figure 5.26 shows a visualisation of vagueness in the definition of boundaries (Davis and Keller, 1997).

Regarding the visualisation of error, MacEachren (1992, 1995) has reviewed the use of changes in geometric attributes such as the visualisation of uncertain values as heights, the application of shading, and colour transformations. Fisher (1994) has proposed the use of sound (see Section 5.7). Fisher (1994), Davis and Keller (1997), and Ehlschlaeger *et al.* (1997) have shown how animation can be used to visualise uncertainty. The latter work illustrates the explanatory strength

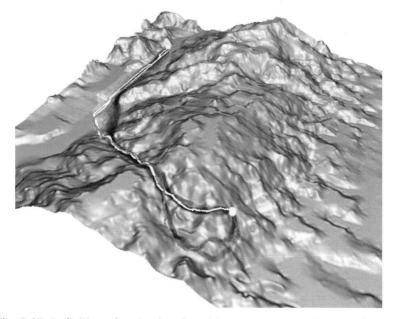

Fig. 5.27 Definition of optimal paths subject to uncertain elevation data. All optimal path realisations are represented. The book's Web site shows the animated visualisation

Reprinted from *Computers & Geosciences*, Vol. 23, C.R. Ehlschlaeger, A.M. Shortridge, and M.F. Goodchild, Visualizing Spatial Data Uncertainty, pp. 387–95. Copyright 1997, with permission from Elsevier Science

of animation: on the definition of optimal paths subject to uncertain elevation data, hundreds of realisations are generated (Fig. 5.27). Subsequent animations of the generated surfaces using interpolation between the surface realisation enable the comprehension of the effect of uncertainty.

5.6 Virtual Environments

The development of virtual environments consists of modelling and visualising real or imaginary worlds in three dimensions and in real time. The emergence of VRML (Virtual Reality Modelling Language) now offers the possibility of developing virtual environments at a lower cost and making them available to a wide audience in the Internet. This means that people anywhere in the world could participate in the solution of environmental problems.

Two other virtual environment technologies are augmented reality (AR) and telepresence. In AR, one adds virtual information of objects which do not belong to the original scene. Telepresence is associated with the use of remote controls or vehicles that transmit images and enable the manipulation of objects.

5.6.1 Virtual Reality

Camara *et al.* (1998) argued that site investigations in environmental studies are reality based and provide sensory immersion but only allow limited exploration in space and time dimensions. Traditional interactive computer systems, despite their graphic interfaces, rely on symbolic computations, do not provide sensory immersion, and allow analysis along the space and time dimension in a compressed period of time.

Virtual reality (VR) technologies, providing the real-time generation of quasi-realistic three-dimensional graphics and sound, combine the advantages of realism of site investigation with the 'what . . . if?' capabilities of interactive systems. In addition, they provide sensory immersion.

Virtual worlds or virtual environments (VEs) facilitate human–computer interaction with environmental decision support systems by the use of 'realistic' representations and direct manipulation of virtual objects. Virtual Tejo, a system developed by Camara *et al.* (1998) and discussed in Chapter 7, is an illustrative example. They can also be used in three-dimensional visualisation of environmental data.

Applications of virtual environments to environmental quality problems include:

- Wheless *et al.* (1996), with visualisations of a water quality model for the Chesapeake Bay.
- Gaither *et al.* (1997), with the visualisation of ocean circulation models.
- Camara *et al.* (1998), with a decision support system for water quality management with a VR front-end.
- Cook *et al.* (1998), with the exploration of environmental data in a CAVE environment (see Section 5.3.1 for description of the work and below for an introduction to CAVE).

Virtual reality has also been applied to the interactive visualisation of artificial ecosystems as described in Heudin (1999). Nerve Garden (see Chapter 4) and Virtual Great Barrier Reef (covered in Chapter 6) are illustrative examples discussed in Heudin's book. There have also been VRML applications related to terrain visualisation, which are presented below.

Fundamental Concepts

Virtual environments provide the real-time generation of quasi-realistic three-dimensional graphics and sound and enable object manipulation in real time (Burdea and Coiffet, 1994). In traditional VR, the user is interacting in a three-dimensional world generated by a computer using appropriate displays and position/orientation sensors.

The three-dimensional perspective results from the slightly different versions that each of our eyes has of a scene. This binocular disparity provides the mechanism for seeing three-dimensional information. There are now several computer displays and head-mounted displays that show three-dimensional images (Jones, 1996; Brice, 1997). They fall under non-immersive and immersive categories. Non-immersive solutions include the use of glasses where the lenses consist of fast shutters synchronised to the computer display. The shutters, using liquid crystal display (LCD) technology, alternately display different images for each eye (Jones, 1996).

The head-mounted displays (HMDs) are popular immersive approaches. HMDs are usually helmets containing two displays, one for each eye, with the images filling the user's optical field. Coupled with devices to track the position and orientation of the user's head (Meyer *et al.*, 1991; McKenna and Zelter, 1992), they give the computer the parameters to render the images and sounds in real time.

The Cave Automatic Virtual Environment (CAVE) (Cruz-Neira *et al.*, 1992) is based on a projection system (the user is inside a system defined by projected images). It is not fully immersive but enables the sharing of virtual environment by groups of people (Fig. 5.28).

The Immersadesk (Reed *et al.*, 1997) supporting interaction with a large graphical surface viewable in stereo three-dimensions using LCD eyepieces is another system used to visualise virtual environments (Fig. 5.29).

Holograms are yet another device for providing three-dimensional images. Holograms are images generated by the use of light waves produced by a laser when this emits on an object and which are then captured by high-resolution

Fig. 5.28 The CAVE environment (a cube with 8 inch sides)

Source: http://www.ncsa.uiuc.edu/Vis

Fig. 5.29 The Immersadesk environment
Source: http://www.ncsa.uiuc.edu/Vis

film. The images look three-dimensional because of the light patterns as they go through the film (Stuart, 1996). A hologram stores not a single image, but in essence a series of images, one for each possible viewing direction (Jones, 1996). MIT Media Lab's Spatial Imaging Group (see the Web site address in Section 5.8) has developed auto-stereoscopic displays based on interactive video holography. Currently, they only use one colour, have a small viewing area, and have considerable computational requirements.

Objects in virtual environments have a hierarchical structure: nodes group several objects, objects group several polygons, and polygons include several vertices. Vertices are the atomic elements described by their (x, y, z) coordinates. The objects that populate a virtual environment can be produced by defining the vertices and polygons in a text editor, using a three-dimensional modelling package or working directly from real objects applying a three-dimensional digitising system. They can also be produced in real time in the virtual environment as a result of user modelling or as an output of a simulation process (Stuart, 1996). Objects may be defined by attributes such as:

- position and orientation;
- scale;
- colour, texture, and shape;
- visibility;
- attached sound;

- behavioural properties;
- interaction relationships with other objects.

A key characteristic of virtual reality systems is latency. Latency is the accumulation of lag introduced by each system component required to generate the system response as a result of (Stuart, 1996):

- The data rate and update rate of position trackers.
- The time spent by the computer in processing the tracker position, running the simulation, and rendering the graphical and sonic images.
- The refresh rate of display devices and the cumulative transmission time.

Virtual reality enables one to fly, walk, and dive into representations of natural systems, select any point of view, manipulate objects, and scale them. (Web addresses of software to develop virtual worlds are available in Section 5.8.)

Virtual Ecosystems

Dias *et al.* (1995) and Almada *et al.* (1996) explored virtual reality capabilities to develop the concept of virtual ecosystems. In a virtual ecosystem:

- there is a background of air, land, or water, described by three-dimensional models draped with appropriate textures (usually photo-textures). There is an associated environmental quality database;
- there is a collection of artificial entities, including polluting and pollutant-removal systems, represented by three-dimensional quasi-realistic models with an associated database;
- there is a collection of natural entities, including fauna and flora, also represented by quasi-realistic models and associated with a database;
- there is a set of transition rules representing the behaviour and interaction among the background, and artificial and natural entities. These transition rules may be represented in traditional environmental quality models or using cellular automata and other individual based approaches.
- One can manipulate entities and change the background. These activities activate the environmental models.
- The results of these models may be visualised in the virtual environment from any point of view.

From a visualisation standpoint, this concept relies on the real-time rendering of large backgrounds, such as terrains, and on the visualisation of ecosystems and their stresses on the virtual environment.

Visualisation of Large Terrains

To visualise large terrains, Muchaxo *et al.* (1999) developed a method that begins by considering a digital terrain model draped with aerial photographs or satellite images. The digital terrain model is represented using a triangulated irregular network (TIN). The rendering requirements of large terrains indicates the adoption of a level-of-detail (LOD) management approach. A LOD implies that the rendering detail in the terrain should be inversely proportional to the distance of the viewpoint.

LODs are based on partitioning the terrain data following methods such as the ones proposed by Muchaxo *et al.* (1999). These implementations enable the generation of images according to pre-established error boundaries for geometry (related to the digital terrain model) and texture detail (related to the phototextures draping the terrain model). The approach used by Muchaxo *et al.* (1999) was to design a tool that subdivides a terrain hierarchically into four tiles (or quad cells) and that generates TINs for each quad cell. The process is repeated for each of the subterrains until a maximum recursion depth is achieved. A quadtree is assembled in the process, resulting in several LODs for each level of the tree.

The TINs are created using a modified version of the DeFloriani's triangulation (DeFloriani *et al.*, 1985). DeFloriani's triangulation consists of starting with a coarse regular triangulation and breaking each polygon in tree polygons using an 'inner' point. That point is the surface point which deviates the furthest from the triangle. The subdivision is repeated until the triangulation is within a given tolerance. That tolerance is a function of the quadtree depth. The selected point can also be over the edge, allowing the triangle to be broken into two triangles.

The Muchaxo *et al.* (1999) algorithm also takes into account terrain features by simply giving more relevance to the deviations over the edges. This method aims to capture the terrain's critical lines such as valleys, ridges, or channels.

Along with the creation of the meshes, the large texture that one wants to map on the terrain is also broken into several tiles according to the quadtree subdivision. One can then create textures with different levels of detail for each level: four pixels at the quadtree depth $N + 1$ (fine level) are averaged into one pixel in the quadtree depth N (coarser level). Quadtrees have to be created from 'leaves' to 'root'. One has to start at the finer level (the leaves of the quadtree depth M), create triangulations and break the texture into tiles, and only then go to depth $M-1$; where one quad cell corresponds to four quad cells on level M).

In real time, according to the distance to the viewer, one simply has to go down the quadtree (root to leaves) and select the proper quad cell to be rendered. As a result, the scene generated is such that it provides decay in polygon size and in texture detail from the viewpoint position to infinity.

The book's Web site includes a demonstration of the application of this method to a fly-over across Laguna Beach, California. This work was implemented using World Toolkit on top of Open GL (see Section 5.8 for the Web addresses). Figure 5.30 shows the underlying terrain models used in Digital Portugal (see Box 5.1), a successful implementation of this algorithm.

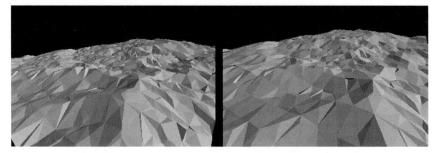

Fig. 5.30 Terrain models underlying Digital Portugal

The LOD technique presents a critical problem if one wants the ecosystem background to be changed over time: any change in a level has to be immediately updated in all other levels. This was impossible to achieve in a real-time generated environment, with the machines used in this study.

Muchaxo and Neves (1999) developed a method to handle the different resolution levels in the visualisation in real time of large terrains using wavelet functions (see Appendix 2). In simple terms, Muchaxo and Neves proposed that instead of changing geometry, one only has to change coefficients that represent the different resolution (or detail) levels. Any change at a given level can then be updated instantly at all other levels. Gross *et al.* (1996) developed a related work by using wavelets and quadtrees to model terrains. However, this technique does not check the generation of images according to pre-established error boundaries.

Box 5.1 *Digital Portugal*

In Digital Portugal (Neves *et al.*, 1998), a mosaic with aerial photos, covering Lisbon and Porto with 1 m resolution, and LANDSAT images covering the rest of the country at 25 m resolution, was used. This mosaic draped a digital terrain model provided by Portugal's Instituto Geográfico do Exército.

The users could fly seamlessly over the whole country by manipulating a joystick. In neighbouring screens, 40 thematic databases on the areas being flown over could be browsed. Figure 5.31 shows the Digital Portugal installation and Fig. 5.32 displays a view obtained from a virtual flight. This project was shown at Expo98 and used by hundreds of thousands of visitors.

Digital Portugal's virtual reality component is installed in a Silicon Graphics Onyx2 Infinite Reality. The real-time rendering was carried out at a rate of 20 frames per second. The system will be progressively ported to SNIG, the national infrastructure on spatial information covered in Chapter 3.

Fig. 5.31 The Digital Portugal installation at Expo98

Fig. 5.32 View of a fly-over across Portugal

Visualisation of Water Pollution

Camara *et al.* (1998) displayed visualisations of pollutant particles in virtual environments (see Figs 7.15 and 7.16), by using water quality models for particle generation and tracking, particle rendering and fog effects.

In virtual environments, a user may select any viewpoint. Camara *et al.* (1999) explored this possibility by applying image filtering to emulate animal vision. Simple filters were developed to simulate the vision of animals with frontal eyes (offering good depth perception) and lateral eyes (providing panoramic views). These filters were inspired by the Supersense program presented in Chapter 2 (Downs, 1988).

The methods are based on applying image filtering algorithms to an image stored in memory (image generated from the fish viewpoint) as a two-dimensional array of RGB (red, green, blue). A filtered image is then displayed in a different window as in Fig. 5.33.

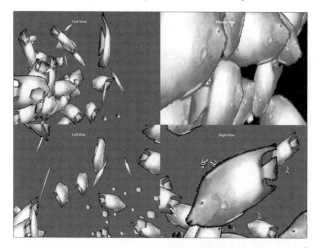

Fig. 5.33 Windows represent fish viewpoints. The upper right window shows the filtered view

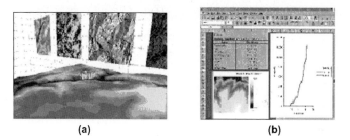

(a)   (b)

Fig. 5.34 Multiple views of a forest fire model: (a) virtual reality view; (b) Excel representation including cellular automata model output

Multiple Views

Gonçalves *et al.* (1997) showed how cellular automata models can be visualised on Excel spreadsheets and simultaneously on virtual environments using the interoperability provided by Microsoft's Component Object Modelling (COM) and a tool called Virtual GIS Room proposed by Neves *et al.* (1997), and further discussed by Neves and Camara (1999) (see Chapter 6 for interface design considerations). In this case, the cellular automata model and the virtual environment run independently but exchange objects with each other.

The Virtual GIS Room application updates the digital terrain model, where a correspondence between cell sizes and the virtual environment and the simulation environment is established when the forest fire model starts running. The user can, at the same time, change model inputs on an Excel spreadsheet. Figure 5.34 displays the multiple views offered by this arrangement.

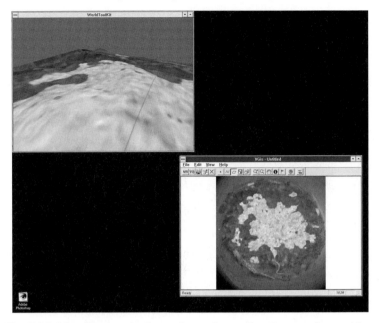

Fig. 5.35 Visualisation, in three dimensions, of actions performed in a two-dimensional space

The Virtual GIS concept is in itself an example of multiple views for the manipulation of geographical information. One can edit two-dimensional representations and visualise the impacts on a three-dimensional real-time generated image, as shown in Fig. 5. 35.

Immersive Photography

Apple's QuickTime VR and the IPIX systems, introduced in Chapter 2, provide an inexpensive method to explore photos wrapping around a limited area. They can be readily published on the Internet as shown on the companies' Web sites (Section 5.8). They do not provide the same general VR properties, such as selecting any point of view or scaling and object manipulation, but they enable a quasi-three-dimensional exploration of spaces such as The Monumental Core Virtual Streetscape site developed at MIT (see the Web address in Section 5.8).

5.6.2 Internet Applications

Vince and Earnshaw (1999) provide a recent review of virtual environments on the World Wide Web where 'VRML' is the key term. VRML is a file format for describing three-dimensional objects and worlds on the Web. It developed from Sillicon Graphics' object-oriented Open Inventor three-dimensional toolkit.

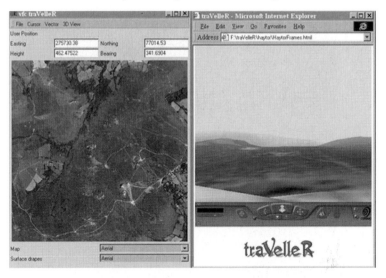

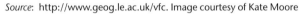

Fig. 5.36 The traVelleR screen, a VRML–Java enabled tool for exploring three-dimensional spaces

Source: http://www.geog.le.ac.uk/vfc. Image courtesy of Kate Moore

VRML worlds can use formats, such as JPEG, MPEG, and MIDI, by way of the Multipurpose Internet Mail Extensions (MIME). These media formats can be scripted within VRML using Java and Java Script and externally by using a Java application (Naik, 1998). VRML worlds can be developed using authoring tools or programmed directly.

VRML 2.0 (Ames *et al.*, 1997) has been the basis for applications to geography and environmental systems. They include:

- The use of VRML in cartography by Fairbairn and Parsley (1997).
- Rhyne's work in the interface of GIS and VRML (see EPA's Virtual GIS site in Section 5.8).
- Djurcilov and Pang (1998) interactive visualisations in VRML of meteorological and environmental phenomena.
- The Virtual Field Course project led by three UK University groups (see Web site listed in Section 5.8). Figure 5.36 illustrates a Java–VRML tool developed in this project. Figure 5.37 shows a screen of LandSerf, a related Java based project.
- TerraVision II, a browser for massive terrains which uses Java scripting to extend VRML base capabilities and an External Authoring Interface for managing specific applications in the virtual geographical environment. TerraVision II has been designed to enable browsing of the earth. (Reddy *et al.*, 1999).

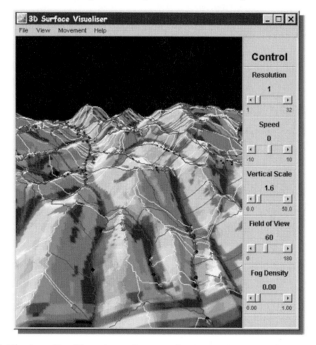

Fig. 5.37 The LandSerf Java based screen for raster images of terrains overlaid with vector representations

Source: http://www.geog.le.ac.uk/vfc. Image courtesy of Jo Wood

VRML 2.0 has now been updated to VRML 97: this world is essentially a hierarchic graph with nodes that are event generators and others that are event receivers. Most event generator nodes are sensor nodes that enable the interactivity in VRML (by detecting touching, proximity, collisions, and visibility). Other event generator nodes may be script nodes that can be defined by Java or JavaScript to trigger events in the VRML world. A sample of relevant VRML nodes is displayed in Table 5.1.

A VRML repository may be found in the Web3D consortium site (Section 5.8). At this site there is updated information on expected VRML developments such as providing database functionality within VRML. The use of VRML for cartographic presentation is debated by the GeoVRML group (see the Web address in Section 5.8).

5.6.3 Augmented Reality

By superimposing synthetic on to real images in the view field of a head-mounted display (HMD) with see-through capabilities, reality is augmented (Robinett, 1992; Feiner *et al.*, 1993*a*,*b*, 1997). There are applications of aug-

Table 5.1 Sample of relevant VRML nodes

Anchor	Fog	Sound
Appearance	Group	Sphere
AudioClip	LOD	SphereSensor
Background	Material	SpotLight
Box	Movie Texture	Switch
Collision	NavigationInfo	Text
Colour	PixelTexture	TextureCoordinate
Cone	PlaneSensor	TimeSensor
Coordinate	PointLight	TouchSensor
Cylinder	ProximitySensor	Viewpoint
Directional Light	Script	VisibilitySensor
ElevationGrid	Shape	WorldInfo

Source: Ames *et al.*, 1997.

mented reality (AR) in manufacturing (labels and diagrams superimposed on to real images) as pointed out by Robinett (1992), and in many other fields as reviewed in the site maintained by Jim Vallino (see the Web site address in Section 5.8). An urban related application has been developed by Feiner *et al.* (1997) to associate labels to buildings at Columbia University. A geographical positioning system (GPS) was used to position the user's display (Figs 5.38 and 5.39). The same researchers intend to generate an augmented reality of the tunnels underneath the university again using GPS.

Barfield *et al.* (1995) propose augmented reality methods to also include environments where video is used to capture the real world. Displays may either be opaque HMDs or any other screen-based system.

Augmented reality has several advantages (Barfield *et al.*, 1995):

- The computational resources required to add synthetic imagery to real-world scenes are much less than the ones demanded to generate those scenes.
- By using real-world scenes, a high level of detail and realistic shading are maintained.
- Augmented reality displays minimise simulator nausea, because the observer still has an anchor in reality.

The main problems associated to augmented reality are (Barfield *et al.*, 1995):

- Image registration, or positioning of the synthetic objects in relation to the real objects.
- Optics of the camera lens used to acquire real-world images and the optics of the HMD. The field of view of the camera lens and the HMD's optics must match exactly, otherwise, magnification or 'minification' of the real-world view will occur.

Fig. 5.38 The Touring Machine: backpack, head-mounted display, hand-held display, and a stylus

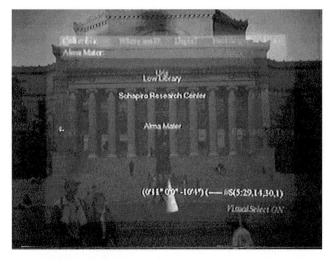

Fig. 5.39 The Touring Machine see-through display: the selected target is the Alma Mater statue; the labels refer to buildings in a straight line behind Low Library at Columbia University

- Issues such as the frame rate, update rate, system delays, and the sensitivity of tracking sensors.

Other difficulties common to wearable AR systems, such as wireless Internet access and geo-referencing, are discussed in Chapters 6 and 8, respectively.

5.6.4 Telepresence

Telepresence is generally associated with the remote operation of vehicles or robots that transmit images and enable the manipulation of objects under harsh conditions such as in Mars exploration (McGreevy, 1993), undersea environments, nuclear reactors, and hazardous waste disposal sites. Telepresence uses devices, such as head-mounted displays and force-feedback handgrip, that are linked to the distant vehicle or robot that have video cameras. These cameras turn with the operator's head and the robot arm moves according to the operator's hand motions (Robinett, 1992).

Microteloperation replaces the robot with a microscope and micromanipulator. Robinett *et al.* (1992) developed a system using a scanning–tunnelling microscope (STM) that uses a tiny probe. This probe captures a three-dimensional image of the surface (at atomic resolution) and can be used as micromanipulator to interact with the sample material. Applications of this system to the micromanipulation of microbial films in sanitary engineering applications could be challenging.

5.7 The Use of Hearing and other Senses

5.7.1 Hearing

Vision depends on the reflection of light from surfaces. Vibrating materials cause sound. Gaver (1989) pointed out that, as a result, sound can provide information that vision cannot such as about occluded materials or the internal mechanisms of complex objects.

The use of sound in visualisation implies (Kramer, 1994; Barrass and Kramer, 1999):

- The association of sounds to data points.
- The control of the sound attributes by data values.
- The triggering of the sound on some event.

For sonification exercises, the National Center for Supercomputing Applications has available now the NCSA Vanilla Sound Server (see NCSA Web site in Section 5.8).

The use of sound in environmental visualisation (or auralisation) complements the role of images as (Shepherd, 1994):

- Multidimensional data can be interpreted because there are several sound parameters and each one can be used for each dimension.
- Parameters that change in time can be represented.
- Outliers and extreme values can be highlighted.

In an animation of the evolution of Yellowstone National Park, Scaletti and Craig (1993) provided examples of the use of sound in the visualisation of environmental phenomena. Shiffer (1993) has also used sound to demonstrate the impact of airplanes in taking off from airports in a multimedia application. In another example, Shiffer (1993) has linked a video on urban traffic with the corresponding noise levels in a planning support system for a central area in Washington, DC. Krygier (1994) reviewed other spatial but non-environmentally related uses of sound in spatial problems.

Wenzel *et al.* (1990) described the techniques used to generate three-dimensional sounds. Those techniques involve the mapping of the qualities of sounds as they reach the ears from many directions. In pioneer experiments, Wenzel and colleagues moved sound sources to various points in a room and made measurements. From these data, head-related transfer functions (HRTFs) were created to modulate the sounds as the sound source moved relative to the user's head. HRTFs also accounted for changes in orientation and location of the user's head, measured by position sensors, and this indicated that it was possible to add three-dimensional sound to virtual environments.

Sound can then help the user locate sources of information which are outside of the field of vision in virtual environments. In these environments, it is also recommended to vary the location of the sound sources horizontally, as vertical judgement is less accurate (Begault, 1994; Hereford and Winn, 1994).

Sound becomes a more significant guiding factor than visual variables when immersed in a virtual aquatic environment according to Neves *et al.* (1994). There, the auralisation of pollutant levels can use surround sound to represent the global water pollution level at a given place, and localised sound to guide users to the most significant concentrations of pollutant particles. The use of sound worked, in that case, in parallel with the graphic rendering, without any decrease in overall performance.

Ellis (1998) and Berry (1999) discuss the use of sound in VRML environments. VRML2 deals with sound through the Audio Clip node and the Sound node. The Audio Clip node may be used to retrieve and play a sound file from any Universe Resource Locator. Using the Audio Clip node as a source, the Sound Node enables the spatialisation of sound. To determine the extent and shape of the sound, this node allows the specification of the direction and of how far in front and behind the sound extends.

The use of sound in visualisation efforts still has unsolved problems, such as the lack of models (unlike colour), and the choice of perceptual auditory parameters (i.e., pitch, intensity, timbre). The modelling of different sound sources is another difficult problem common to noise assessment studies.

5.7.2 Other Senses

Robinett (1992) and Holloway and Lastra (1993) discussed the alternative methods of providing synthetic experiences involving senses other than vision and hearing. For environmental systems, the most relevant are haptic senses (kinaesthetic, tactile senses), smelling, and tasting. There has been some limited success in synthesising these senses in digital systems. The exceptions are the haptic senses, where the videogame industry has developed inexpensive force-feedback devices. These follow the pioneering work of Brooks *et al.* (1990). Current systems, such as Phantom from Sensable Technologies (see Web site in Section 5.8), have already such capabilities as enabling one to feel shapes and textures and modulate and re-shape objects. In addition, these capabilities can be programmed. The Sensorama experience (Krueger, 1983, 1991) provided an analogue alternative to generate smells. Digiscents has recently introduced a digital approach with its iSmell technology (see Section 5.8 for the Web site). Robinett (1992) has discussed the generation of taste from four dimensions (salty, sour, sweet, and bitter, which are related to chemical content), just as primary colours are synthesised from the red, green, and blue dots of the television screen.

The simulation of sensory experiences should be particularly relevant in landscape analysis, where the visual and sound features could be complemented by smell, taste, and touch.

5.8 Further Exploration

The following sites provide extensive information and tools on the topics covered in this chapter. The list is updated at (http://gasa.dcea.fct.unl.pt/camara/index.html). Student projects are also proposed at this site.

Typography
Studiomotiv's site offers an excellent tutorial on typography (http://www.studiomotiv.com/counterspace).

Statistical Graphics
General Web sites in statistical graphics are offered by:

- The Statistical Graphics section of the American Statistical Association (ASA) (http://www.bell-labs.com/topic/societies/asagraphics). The ASA produced a video on Statistical Graphics that synthesises the major developments in the field.
- Eugene Horber with a site on Exploratory Data Analysis (http://www.unige.ch/ses/sococ/eda), and Dianne Cook's links on Statistical Graphics (http://www.public.iastate.edu/~dicook).

Commercial software with statistical graphics capabilities include:

- Systat and SPSS/Diamond software (http://www.spss.com).
- SAS/Insight software (http://www.sas.com).
- Datadesk software (http://www.datadesk.com).

X-Gobi is software for high dimensionality data visualisation that can be freely downloaded (http://www.research.att.com/areas/stat/xgobi).

Other public domain statistical software includes Globally Accessible Statistical Procedures (GASP) (http://www.stat.sc.edu/rsrch/gasp) and the Statistical Library (http://lib.stat.cmu.edu).

Introductions to data mining software using visual methods are available at the following sites:

- Spotfire software (http://www.spotfire.com).
- Silicon Graphics' Mineset (http://www.sgi.com/software/mineset).
- Cognos' Power Play (http://www.cognos.com).

For applications of statistical graphics to environmental problems see Dianne Cook's site (Universe Resource Locator above). The project using CAVE is described at (http://www.public.iastate.edu/~dicook/research/c2/statistic.html).

The 'worlds-within-worlds' application is available at (http://www.cs.columbia.edu/graphics).

Scientific Visualisation

General sites on visualisation and animation include:

- OpenGL, the foundation for high performance graphics, is available at (http://www.opengl.org).
- ACM SIGGRAPH, the organiser of the major computer graphics conference (http://www.siggraph.org). SIGGRAPH videos are demonstrative of the developments in these areas.
- ACM SIGGRAPH Online Bibliography Database, with more than 16.000 publication in computer graphics (http://www.cgrg.ohio-state.edu/~spencer/newbib.html).

- A commercial site covering 3D computer graphics is 3D Site (www.3dsite.com).
- Links on scientific visualisation projects, bibliography, and software produced by centres located around the world are available at (http://www.nas.nasa.gov/Groups/VisTech/VisWeblets.html).
- National Center for Supercomputing Applications Web site on visualisation (http://www.ncsa.uiuc.edu/Vis).
- Chris Jones' site on visualisation and optimisation (http://www.chesapeake2.com/itorms).
- Mining Company list of links on animation (http://animation.miningco.com/mbody.htm).
- Craig Reynolds reviews procedural animation (http://www.red.com/cwr/behave.html).
- University of California, Santa Cruz Laboratory for Visualisation and Graphics reviews visualisation of uncertainty (http://www.cse.ucsc.edu/research/slvg).

Commercial software for scientific visualisation are introduced at the following sites:

- Silicon Graphics software (http://www.sgi.com).
- IBM Visualisation Data Explorer (http://www.ibm.com).
- Research Systems software including NOESYS (general purpose visualisation tools), Interactive Data Language for visualisation, ENVI (an ENVIronment for visualising satellite images), and River Tools (http://www.rsinc.com).
- WSI software for visualisation in weather forecasting (and broadcasting) (http://www.wsicorp.com).
- Environmental Visualisation Systems software for visualisation in mining and geology related applications (http://www.ctech.com).

Public domain software for scientific visualisation is available at:

- Master Engineering Library (http://mel.dmso.mil). This site enables the downloading of VIS 5D (data visualisation in three dimensions and animation), Interactive Data Language (IDL), and FERRET, a software for visualisation in oceanography and meteorology.
- Jet Propulsion Laboratory's Linkwinds site (http://linkwinds.jpl.nasa.gov), Linkwinds is a collaborative visualisation system.

Commercial software and services for visualising terrains and landscapes are, in most cases, Web enabled (they generate QuickTime VR panoramas and VRML files). Information on some of the existing products is available at:

- ESRI's site (http://www.esri.com). This site introduces ArcView 3D Analyst among other relevant applications.
- Rapid Imaging System's Landform site (http://landform.com).
- Geomantics' site (http://www.geomantics.co.uk).
- Formz's site (http://www.formz.com).
- ErMapper's site (http://www.ermapper.com).
- Bryce4 site (http://www.metacreations.com).

GRASS is public domain software that may be used to visualise terrains as shown at (http://www.baylor.edu./~grass). The University of Toronto's Center for Landscape Research makes available tools for landscape visualisation (http://www.clr.toronto.ca).

Leading animation software is introduced at the following sites:

- Macromedia (Director, Flash and Shockwave) (http://www.macromedia.com).
- 3D Studio Max (http://www.ktx.com).
- Strata 3D (http://www.strata3d.com).
- Alias Research products such as MAYA (http://www.aw.sgi.com).
- Softimage (http://www.softimage.com).

For a site with links for freeware or shareware on animation see (http://www-personal.umich.edu/~jeffab/graphics.html). Innovative applications of interactive animation are provided at (http://www.freestyleinteractive.com/demos/).

Environmental applications of visualisation and animation software are available at the following sites:

- EPA's Visualisation Laboratory (http://www.epa.gov/vislab/svc/index. html). At this site see also the visualisation of the next generation air quality models (http://www.epa.gov/vislab/svc/projects/models3/index.html).
- Woods Hole site for marine data visualisation (http://woodshole.er.usgs.gov).
- University of California, Santa Cruz Laboratory for Visualisation and Graphics work on environmental data visualisation (http://www. cse.ucsc.edu/research/slvg).
- Center for Landscape Research at University of Toronto (http://www.clr.utoronto.ca).
- International Cartographic Association's visualisation efforts (http://www.geovista.psu.edu/ica/ICAvis).
- Geo@News project at the New University of Lisbon (http://gasa.dcea.fct.unl.pt/geo@news/index.html).
- Mark Monmonier's Air Apparent book site (http://www.press.uchicago. edu) includes visualisation methods used in weather forecasting.

- Internet 1996 World Exposition site (http://park.org) includes Japan's The Breathing Earth experiment.
- Craig Reynolds' Boids Java animation (http://www.red.com/cwr/behave.html).
- Artificial life animation tools (http://www.ventrella.com).
- Java based animations (http://www.euroweather.net/maps/ready.htm).
- Flash animation for environmental education (http://www.secretsatsea.org) and (http://www.planetark.org.)

Virtual Environments

The Human Interface Technology Laboratory of the University of Washington maintains the most comprehensive site on virtual reality (http://www.hitl. washington.edu).

Steve Bryson's lectures in virtual reality, available at (http://science.nas.nasa. gov/~bryson) also provide a general view on virtual reality technologies and applications.

The state-of-the art in interactive video holography is presented at MIT's Spatial Imaging site (http://spi.www.media.mit.edu/groups/spi).

The VRML repository of information and links to applications (as well as information related to Java 3D) is at (http://www.web3d.org). An unofficial repository for materials on augmented reality is maintained by Jim Vallino (http://www.cs.rit.edu/~jrv/research/ar). Another augmented reality repository may be found at (http://www.augmented-reality.org/).

For an introductory site on telerobotics see (http://telerobot.mech.uwa. edu.au).

Commercial virtual reality software solutions are provided by, among others:

- Sense8 (http://www.sense8.com).
- Superscape (http://www.superscape.com).
- Argus (http://www.argusvr.com).
- Fakespace (http://www.fakespace.com).
- Virtex (http://www.virtex.com).

Alice is non-commercial interactive 3D graphics software maintained at Carnegie Mellon, and available for download at (http://www.alice.org). From the same group a manual on how to make a CUBE at home is available at (http://etc.hcii.cs.cmu.edu/projects/cube/equipment.html).

Immersive photography solutions are offered by Apple's QuickTimeVR (http://www.apple.com/quicktime) and IPIX (http://www.ipix.com). An illustrative application is MIT's Monumental Core Virtual Streetscape (http://yerkes.mit.edu/ncpc96/home.html).

Virtual reality applications to environmental and urban planning problems are presented at:

- EPA's Virtual GIS project (http://www.epa.gov/gisvis).
- Virtual Field Course project (http://www.geog.le.ac.uk/vfc).
- SIGGRAPH Carto Project (http://www.siggraph.org/~rhyne/carto) with links for virtual spatial information systems projects around the world and the GeoVRML group. This group is developing standards to solve geographical representation in VRML. Their site is at (http://www.ai.sri.com/geovrml).
- New University of Lisbon's virtual reality projects (http://gasa.dcea.fct.unl.pt/gasa/gasa98.htm).
- Georgia Tech's work on virtual zoo habitats (http://www.cc.gatech.edu/gvu/virtual).
- TerraVision project (http://www.ai.sri.com/TerraVision/TV-II/index.html).
- Nerve Garden project (http://www.biota.org/nervegarden).
- Santa Cruz's VRML visualisations of environmental and meteorological data (http://www.cse.ucsc.edu/research/slvg).
- Center for Coastal Physical Oceanography at Old Dominion University (http://www.ccpo.odu.edu/~wheless) including a VRML visualisation of Chesapeake Bay.
- Virtual Heritage Network (http://www.vsmm.org).
- VRML visualisations of Earth Science projects (http://www.edcenter.sdsu.edu/repository/nav_vrmllinks.shtml).
- MIT's use of QuickTime VR movies in the exploration of Washington, DC (http://yerkes.mit.edu/ncpc96/home.html).
- Columbia University research on augmented reality (http://www.cs.columbia.edu/graphics).
- University of Michican project on augmented reality and detection of hazardous waste (http://www-vrl.umich.edu/sel_prj/ar/hazard).

Sound

General sites on the use of sound in visualisation include Accoustic Society of America's site (http://asa.aip.org) and the World Forum for Accoustic Ecology (http://interact.uoregon.edu/MediaLit/WFAEHomePAge).

The use of sound in VRML is covered at (http://www.dform.com/inquiry/tutorials/vrmlaudio) and (http://www.dform.com/inquiry/spataudio.html).

For software to record and edit soundtracks see Digidesign's site at (http://www.digidesign.com) and Macromedia (http://www.macromedia.com), where the use of BIAS Peak LE (Macintosh) and Sonic Foundry SoundForce XP (Windows) are discussed.

Maria Joao Silva discusses the use of sound in environmental education (http://gasa.dcea.fct.unl.pt/gasa/gasa98.htm).

Haptics

The general site of the haptics community is located at (http://haptic.mech. nwu.edu). This site has a gallery of all the haptic solutions that have been developed around the world. Commercial products include SensAble Technologies (http://www.sensable.com) and Haptech (http://www.immersion.com).

Odour

Digiscents has been a pioneer in bringing smell to the Internet (http:// digiscents.com).

References

Ahlberg, C. and Shneiderman, B. (1994). Visual Information Seeking: Tight Coupling of Dynamic Query Filters with Starfield Displays. *Proceedings of CHI '94*, 313–21. New York: ACM Press.

Almada, A., Dias, A., Silva, J.P., Santos, E., Pedrosa, P., and Camara, A.S. (1996). Exploring Virtual Ecosystems. *Proceedings AVI96 Advanced Visual Interfaces*, 166–74, Gubbio, Italy: ACM Press.

Ames, A.L., Nadeau, D.R., and Moreland, J.L. (1997). *VRML 2.0 Sourcebook*. New York: Wiley.

Anselin, L. (1999). Interactive Techniques and Exploratory Spatial Data Analysis. In P.A. Longley, M.F. Goodchild, D.J. Maguire, and D.W. Rhind (eds), *Geographical Information Systems*, 253–66. New York: Wiley.

Baker, M.P. (1994). The KNOWVIS Project: An Experiment in Automating Visualisation. *Decision Support 2001 Conference*, Toronto, Canada.

Barfield, W., Rosenberg, C., and Lotens, W.A. (1995). Augmented-Reality Displays. In W. Barfield and T.A. Furness III (eds), *Virtual Environments and Interface Design*, 473–513. New York: Oxford University Press.

Barrass, S. and Kramer, G. (1999). Using Sonification. *Multimedia Systems*, 7: 23–31.

Becker, R.A. and Cleveland, W.S. (1987). Brushing Scatterplots. *Technometrics*, 29/2: 127–42.

Becker, R.A., Cleveland, W.S., and Wilks, A.R. (1987). Dynamic Graphics for Data Analysis. *Statistical Science*, 2/4: 355–95.

Begault, D.R. (1994). *3D Sound for Virtual Reality and Multimedia*. Cambridge, MA: Academic Press.

Berson, A. and Smith, S. (1997). *Data Warehousing, Data Mining and OLAP*. New York: McGraw Hill.

Berry, R. (1999). Feeping creatures. In J. Heudin (ed.), *Virtual Worlds*, 211–28. Reading, MA: Perseus.

Bertin, J. (1981). *Graphics and Graphic Information-Processing*. Berlin: Walter de Gruyter.

Beshers, C. and Feiner, S. (1993). AutoVisual: Rule-based Design of Interactive Multivariate Visualisations. *IEEE Computer Graphics and Applications*, 13/4: 41–9.

Bodin, L. and Levy, L. (1994). Visualisation in Vehicle Routing and Scheduling Problems. *ORSA Journal on Computing*, 6/3: 261–9.

Boice, M. (1992). *How to Create a Toxic Plume Map*, CEC Fact Sheet No. 2. Albany, NY: Citizens Environmental Coalition.

Brice, R (1997). *Multimedia and Virtual Reality Engineering*. Oxford: Newnes.

Brooks, F.P., Ouh-Young, M., Batter, J.J., and Kilpatrick, P.J. (1990). Project GROPE—Haptic Displays for Scientific Visualisation. *Computer Graphics*, 24/4: 177–85.

Burdea, G. and Coiffet, P. (1994). *Virtual Reality Technology*. New York: Wiley.

Burrough, P.A. (1989). Fuzzy Mathematical Methods for Soil Survey and Land Evaluation. *Journal of Soil Science*, 40: 477–92.

Burrough, P.A., MacMillan, R.A., and Deursen, W. van (1992). Fuzzy Classification Methods for Determining Land Suitability from Soil Profile Observations and Topography. *Journal of Soil Science*, 43: 193–210.

Buxton, W. (1989). Introduction to this special issue on nonspeech audio. Human Computer Interaction, 4/1: 1–9.

Camara, A.S., Neves, J.N., Muchaxo, J., Fernandes, J.P., Sousa, I., Nobre, E., *et al.* (1998). Virtual Environments and Water Quality Management. *Journal of Infrastructure Systems, ASCE*, 4/1: 28–36.

Camara, A.S., Neves, J.N., Gonçalves, P., Gomes, J.M., Muchaxo, J., Silva, J.P., *et al.* (1999). *Towards Virtual Ecosystems*. Unpublished report, Environmental Systems Analysis Group, New University of Lisbon, Monte de Caparica, Portugal.

Card, S., Eick, S.G., and Gershon, N. (1997). Information Visualisation. CHI '97 Tutorial, Los Angeles, CA.

Chakraborty, J. and Armstrong, M.P. (1996). Using Geographic Plume Analysis to Assess Community Vulnerability to Hazardous Accidents. *Computers, Environment and Urban Systems*, 19: 341–56.

Chernoff, H. (1971). The Use of Faces to Represent Points in K-Dimensional Space Graphically. *Journal of the American Statistical Association*, 76/376: 757–65.

Cleveland, W.S. (1993). *Visualising Data*. Murray Hill, NJ: AT&T Bell Laboratories.

Cook, D., Symanzik, J., Majure, J.J., and Cressie, N. (1997). Dynamic Graphics in a GIS: More Examples using Linked Software. *Computers & Geosciences*, 23/4: 371–85.

Cook, D., Cruz-Neira, C., and Kohlmeyer, B.D. (1998). Exploring Environmental Data in a Highly Immersive Virtual Reality Environment. *Environmental Monitoring and Assessment*, 51/1–2: 441–50.

Cruz-Neira, C., Sandin, D.J., DeFanti, T., Kenyon, R.V., and Hart, J.C. (1992). The CAVE: Audio Visual Experience Automatic Virtual Environment. *Communications of the ACM*, 35: 65–72.

Davis, T.J. and Keller, C.P. (1997). Modeling and Visualising Multiple Spatial Uncertainties. *Computers & Geosciences*, 23/4: 397–408.

DeFloriani, L., Falcidieno, B., Nagy, C., and Pienovi, C. (1985). Delaunay-Based Representation of Surfaces Defined Over Arbitrarily Shaped Domains. *Computer Vision, Graphics and Image Processing*, 32/1: 127–40.

DeGloria, S.D. (1993). Visualising Soil Behavior. *Geoderma*, 60/1–4: 41–55.

Delmarcelle, T. and Hesselink, L. (1995). A Unified Framework for Flow Visualisation. In R.S. Gallagher (ed.), *Computer Visualisation Graphics Techniques for Scientific and Engineering Analysis*. Boca Raton, FL: CRC Press.

Dias, A.E., Silva, J.P., and Camara, A.S. (1995). BITS: Browsing in Time and Space. In Companion Volume ACM CHI '95, 248–9, Denver, CO: ACM Press.

DiBiase, D., MacEachren, A.M., and Krygier, J.B. (1992). Animation and the Role of Map Design in Scientific Visualisation. *Cartography and Geographical Information Systems*, 19/4: 201–14.

Djurcilov, S. and Pang, A. (1998). Web Visualization of Environmental Data. SPIE Conference on Electronic Imaging, San José, CA (oral presentation).

Downs, E. (1988). *Supersense*. London BBC. (video).

Ehlschlaeger, C.R., Shortridge, A.M., and Goodchild, M.F. (1997). Visualising Spatial Data Uncertainty. *Computers & Geosciences*, 23/4: 387–95.

Ellis, S. (1998). Sound and VRML. *Proceedings of VRML98*, 95–100. Monterey Bay, CA: ACM Press.

Fairbairn, D. and Parsley, S. (1997). The Use of VRML for Cartographic Visualisation. *Computers & Geosciences*, 23/4: 475–82.

Fedra, K. (1994). Integrated Environmental Information and Decision-Support Systems. *IFIP Transactions B*, 16: 269–88.

Feiner, S. and Beshers, C. (1990a). Visualising N-Dimensional Worlds with N-Vision. *Computer Graphics*, 24/2: 37–8.

Feiner, S. and Beshers, C. (1990b). Worlds within Worlds: Metaphors for Exploring n-Dimensional Virtual Worlds. *Proceedings of UIST 90*, 76–83. New York: ACM Press.

Feiner, S., MacIntyre, B., Haupt, M., and Solomon, E. (1993a). Windows on the World: 2D Windows for 3D Augmented Reality. *Proceedings ACM UIST 93*, 145–55. Atlanta, GA: ACM Press.

Feiner, S., Macintyre, B., and Seligmann, D. (1993b). Knowledge-based Augmented Reality. *Communications of the ACM*, 36/7: 53–62.

Feiner, S., MacIntyre, B., Hollerer, T., and Webster, A. (1997). A Touring Machine: Prototyping 3D Mobile Augmented Reality for Exploring the Urban Environment. *Proceedings of an International Symposium on Wearable Computing 97*, 74–81, Cambridge, MA.

Fisher, P. (1994). Animation and Sound for the Visualisation of Uncertain Spatial Information. In H.M. Hearnshaw and D.J. Unwin (eds), *Visualisation in Geographical Information Systems*. Chichester, UK: Wiley.

Fisher, P. (1999). Models of Uncertainty in Spatial Data. In P.A. Longley, M.F. Goodchild, D.J. Maguire, and D.W. Rhind (eds), *Geographical Information Systems*, 191–205. New York: Wiley.

Foley, J.D., VanDam, A., Feiner, S.K., and Hughes, J.F. (1990). *Computer Graphics: Principles and Practice*. Reading, MA: Addison Wesley.

Fuchs, F. (1994). A Visualisation System for Marine Environmental Data. *IFIP Transactions B*, 16: 25–36.

Furnas, G.W. (1986). Generalized Fisheye Views. *Proceedings of ACM CHI '86*, 16–34. New York: ACM Press.

Gaither, K., Moorhead R., Nations, S., and Fox, D. (1997). Visualising Ocean Circulation Models Through Virtual Environments. *IEEE Computer Graphics*, 17/1: 16–19.

Gaver, W.W. (1989). The SonicFinder: An Interface that Uses Auditory Icons. *Human-Computer Interaction*, 4/1: 67–94.

Globus, A. and Raible, E. (1994). 13 Ways to Say Nothing with Scientific Visualisation. *IEEE Computer*, 27/7: 86–8.

Goldstein, E.B. (1999). *Sensation and Perception*. Pacific Grove, CA: Brooks/Cole.

Gonçalves, P.P., Neves, J.N., Silva, J.P., Muchaxo, J., and Camara, A.S. (1997). Interoperability of Geographic Information: From the Spreadsheet to Virtual Environments. *International Conference and Workshop on Interoperating Geographical Information Systems*, NCGIA, Santa Barbara, CA.

Green, M. and Sun, H. (1995). Computer Graphics Modeling for Virtual Environments. In W. Barfield and T.A. Furness III (eds), *Virtual Environments and Advanced Interface Design*, 63–101. New York: Oxford University Press.

Gross, M.H, Staadt, C.G., and Gatti, R. (1996). Efficient Triangular Surface Approximations Using Wavelets and Quadtree Data Structures. *IEEE Transactions on Visualisation and Computer Graphics*, 2/2: 130–43.

Groth, R. (1998). *Data Mining*. Upper Saddle River, NJ: Prentice Hall.

Hereford, J. and Winn, W. (1994). Non-Speech Sound in Human-Computer Interaction: A Review and Design Guidelines. *Journal of Educational Computing Research*, 11/3: 211–33.

Heudin, J.C. (ed.) (1999). *Virtual Worlds*. Reading, MA: Perseus.

Holloway, R. and Lastra, A. (1993). *Virtual Environments*. Technical Report. Chapel Hill, NC: University of North Carolina.

Inselberg, A. and Dimsdale, B. (1987). Parallel Coordinates for Visualising Multi-Dimensional Geometry. *Computer Graphics 1987*, 25–44. Berlin: Springer Verlag.

Inselberg, A. and Dimstale, B. (1994). Multidimensional Lines: 1. Representation. *SIAM Journal of Applied Mathematics*, 5/2: 559–77.

Jones, C. (1996). *Visualisation in Optimization*. Norwell, MA: Kluwer (also http://www.chesapeake2.com/itorms).

Kazafumi K., *et al.* (1989). Three-Dimensional Terrain Modeling and Display for Environmental Assessment. *Computer Graphics*, 23/3: 207–14.

Koussoulakou, A. (1994). Spatial-Temporal Analysis of Urban Air Pollution. In A. MacEachren and D.R.F. Taylor (eds), *Visualisation in Modern Cartography*. Oxford: Pergamon Press.

Kraak, M.J. and Ormeling, F.J. (1996). *Cartography, Visualisation of Spatial Data*. Harlow, UK: Longman.

Kramer, G. (ed.) (1994). *Auditory Display: Sonification, Audification and Auditory Interfaces*. Reading, MA: Addison-Wesley.

Krueger, M. (1983). *Artificial Reality*. Reading, MA: Addison-Wesley.

Krueger, M. (1991). *Artificial Reality: II*, Reading, MA: Addison-Wesley.

Kruse, F.A., Lefkoff, A.B., Boardman, J.W., Heidebrecht, K.B., Shapiro, A.T., Barloon, P.J., *et al.* (1992). The Spectral Image Processing System (SIPS)—Interactive Visualisation and Analysis of Imaging Spectrometer Data. *Proceedings, International Space Year Conference, Earth and Space Science Information Systems*, Pasadena, CA.

Krygier, J.B. (1994). Sound and Geographic Visualisation. In A. MacEachren and D.R.F. Taylor (eds), *Visualisation in Modern Cartography*. Oxford: Pergamon Press.

Lacerda, F.W. (1986). Comparative Advantages of Graphic Versus Numeric Representation of Quantitative Data. PhD Dissertation, Virginia Tech, Blacksburg, VA.

Laybourne, K. (1998). *The Animation Book*. New York: Three Rivers Press.

Liddell, C.M. and Hansen, D. (1997). Visualising Complex Biological Interactions in the Soil Ecosystem. *Journal of Visual Computing and Animation*, 4/1: 3–12.

MacEachren, A.M. (1992). Visualising Uncertain Information. *Cartographic Perspectives*, 13: 10–19.

MacEachren, A.M. (1995). *How Maps Work: Representation, Visualisation and Design*. New York: Guilford Press.

MacEachren, A.M. and Kraak, M.J. (1997). Exploratory Cartographic Visualisation: Advancing the Agenda. *Computers & Geosciences*, 23/4: 335–43.

Magnenat-Thalmann, N. and Thalmann, D. (eds) (1994). *Artificial Life and Virtual Reality*. Chichester, UK: Wiley.

Malamud, C. (1997). *A World's Fair for the Global Village*. Cambridge, MA: MIT Press (companion site at http://park.org).

Marcus, A. (1995). Principles of Effective Visual Communication for Graphical User Interface Design. In R.M. Baecker, J. Grudin, W.A. Buxton, and S. Greenberg (eds), *Readings in Human-Computer Interaction: Toward the Year 2000*. San Francisco: Morgan Kauffman.

McGreevy, M.W. (1993). *Virtual Reality and Planetary Exploration*. Boston: Academic Press.

McKenna, M. and Zelter, D. (1992). Three Dimensional Visual Display Systems for Virtual Environments. *Presence*, 1/4: 421–58.

Meyer, K., Applewhite, H.L., and Biocca, F.A. (1991). The Ultimate Tracker: a Survey of Position Trackers. *Presence*, 1/2: 173–200.

Miller, G.F. (1956). The Magical Number Seven, Plus or Minus Two: Some Limits on Our Capacity to Process Information. *Psychological Review*, 63: 81–96.

Mitas, L., Brown, W.M., and Mitasova, H. (1997). Role of Dynamic Cartography in Simulations of Landscape Processes Based on Multivariate Fields. *Computers & Geosciences*, 23/4: 437–46.

Monmonier, M. (1999). *Air Apparent*. Chicago: University of Chicago Press.

Muchaxo, J., Neves, J.N., and Camara, A.S. (1999). A Real-Time, Level-of-Detail Editable Representation for Phototextured Terrains with Cartographic Coherence. In A.S. Camara and J. Raper (eds), *Spatial Multimedia and Virtual Reality*. London: Taylor & Francis.

Muchaxo, J. and Neves, J.N. (1999). *Visualisation of Large Terrains Using Wavelet Functions*. Unpublished internal report, Imersiva, Monte de Caparica, Portugal.

Naik, D.C. (1998). *Internet Standards and Protocols*. Redmond, WA: Microsoft Press.

Neves, J.N. and Camara, A.S. (1999). Virtual Environments and GIS. In P.A. Longley, M.F. Goodchild, D.J. Maguire, and D.W. Rhind (eds), *Geographical Information Systems*, 557–65. New York: Wiley.

Neves, J.N., Carmona Rodrigues, A., and Camara, A.S. (1994). Virtual Reality and Water Pollution Control. In G. Tsakiris and M. Santos (eds), *Advances in Water Resources Technology and Management*, 133–7. Rotterdam: Balkema.

Neves, J.N., Silva, J.P., Gonçalves, P., Muchaxo, J., Silva, J.M., and Camara, A.S. (1997). Cognitive Spaces and Metaphors: A Solution for Interacting with Spatial Data. *Computers & Geosciences*, 23/4: 483–8.

Neves, J.N., Bento, J., and Gouveia, C. (1998). Digital Portugal. Paper presented at GIS Planet, Lisbon, Portugal.

O'Rourke, M. (1998). *Principle of Three-Dimensional Computer Animation*. New York: Norton.

Owen, S.J., Jones, N.L., and Holland, J.P. (1996). A Comprehensive Modelling Environment for the Simulation of Groundwater Flow and Transport. *Engineering Computation*, 12/3–4: 235–42.

Pang, A.T., Wittendrink, C.M., and Lodha, S.K. (1997). Approaches to Uncertainty Visualisation. *The Visual Computer*, 13/8: 370–90.

Peitgen, H.O., Jurgens, H., and Saupe, D. (1992). *Chaos and Fractals*. New York: Springer Verlag.

Prusinkiewicz, P., Lindenmayer, A., and Hanan, J.S. (1990). *The Algorithmic Beauty of Plants*. New York: Springer Verlag.

Quinlan, J. (1992). *C4.5: Programs for Machine Learning*. Mountain View, CA: Morgan Kauffman.

Realmuto, V.J., Abrams, M.J., and Buongiorno, M.F. (1994). The Use of Multispectral Thermal Infrared Image Data to Estimate The Sulphur-Dioxide Flux from Volcanos—A Case Study from Mount Etna, Sicily. *Journal of Geophysics Research*, 99/B1: 481–8.

Reddy, M., Leclerc, Y., and Iverson, L. (1999). Terra Vision: II. Visualising Massive Terrain Databases in VRML. *IEEE Computer Graphics and Applications*, 19/2: 30–8.

Reed, D.A., Giles, R.C., and Catlett, C.E. (1997). Distributed Data and Immersive Collaboration. *Communications of the ACM*, 40/11: 39–48.

Reeves, W. (1983). Particle Systems—A Technique for Modelling a Class of Fuzzy Objects. *ACM SIGGRAPH 83 Proceedings*, 359–76. New York: ACM Press.

Reynolds, C.W. (1982). Computer Animation with Scripts and Actors. *Computer Graphics*, 16/3: 289–96.

Rhyne, T. (1997). Internetworked 3D Computer Graphics: Beyond Bottlenecks and Roadblocks. Notes for an ACM SIGCOMM 97 tutorial (unpublished).

Robinett, W. (1992). Synthetic Experience: A Proposed Taxonomy. *Presence*, 1/2: 229–47.

Robinett, W., Taylor, R., Chi, R., and Wright, W.V. (1992). The Nanomanipulator: an Atomic-Scale Teleoperator. Technical report. Chapel Hill, NC: Computer Sciences Department, University of North Carolina.

Robinson, V. (1988). Some Implications of Fuzzy Set Theory Applied to Geographical Databases. *Computers, Environment and Urban Systems*, 12: 89–98.

Rogowitz, B.E. and Treinish, L.A. (1994). An Architecture for Perceptual Rule-Based Visualisation. *Proceedings of the IEEE Visualisation '94 Conference*, 236–43. San José, CA: IEEE Computer Society Press.

Rohrer, R. and Swing, E. (1997). Web-Based Information Visualisation. *IEEE Computer Graphics and Applications*, 17/4: 52–9.

Santos, M.P. (1994). Lecture Notes on Scientific Visualisation. PhD Program in Environmental Engineering, New University of Lisbon, Monte de Caparica, Portugal (unpublished).

Scaletti, C. and Craig, A.D. (1993). *Using Sound to Extract Meaning from Complex Data*. Technical Report. Urbana-Champaign: CERL, University of Illinois.

Sheperd, I. (1994). Multi-Sensory GIS: Mapping Out The Research Frontier. *Proceedings Spatial Data Handling 94*, 356–70, Edinburgh, Scotland.

Schroeder, W., Martin, K., and Lorensen, B. (1998). *The Visualisation Toolkit*, 2nd edn. Upper Saddle River, NJ: Prentice Hall.

Shiffer, M. (1993). Augmenting Geographic Information with Collaborative Multimedia Technologies. *Proceedings of AUTOCARTO 11*, 367–76, Minneapolis, MN.

Shneiderman, B. (1996). *The Eyes Have It: A Task by Data Type Taxonomy for Information Visualisations*. Technical Report. College Park, MD: University of Maryland.

Sims, K. (1990). Particle Animation and Rendering using Data Parallel Computation. *ACM SIGGRAPH 90 Proceedings*, 405–13. New York: ACM Press.

Spiekermann, E. and Ginger, E.M. (1993). *Stop Stealing Sheep and Find Out How Type Works*. Indianapolis, IN: Adobe Press.

Stuart, R. (1996). *The Design of Virtual Environments*. New York: McGraw-Hill.

Terzopoulos, D., Tu, X., and Grzeszczuk, R. (1994). Artificial Fishes with Autonomous Locomotion, Perception, Behavior, and Learning in a Simulated Physical World. In R. Brooks and P. Maes (eds), *Artificial Life: IV*. Cambridge, MA: MIT Press.

Thomas, F. and Johnston, O. (1984). *Disney Animation. The Illusion of Life*. New York: Abbeville Press.

Tufte, E. (1983). *The Visual Display of Quantitative Information*. Cheshire, CT: Graphics Press.

Tufte, E.R. (1990). *Envisioning Information*. Cheshire, CT: Graphics Press.

Tukey, J. (1977). *Exploratory Data Analysis*. Reading, MA: Addison-Wesley.

Vince, J. and Earnshaw, R. (1999). *Virtual Worlds on the Internet*. New York: IEEE Press.

Wade, N.J. and Swanston, M. (1991). *Visual Perception: An Introduction*. London: Routledge.

Wagstaff, S. (1999). *Animation on the Web*. Berkeley, CA: Peachpit Press.

Wenzel, E.M., Stone, P.K., Fisher, S.S., and Foster, S.H. (1990). A System for Three-Dimensional Accoustic Viewing in a Virtual Environment Workstation. *Proceedings of the First IEEE Conference on Visualisation*. Los Alamitos, CA: IEEE Computer Society Press.

Wheless *et al.* (1996). Virtual Cheasapeake Bay: Interacting with a Coupled Physical/Biological Model. *IEEE Computer Graphics and Applications*, 16/4: 52–7.

Wittenbrink, C.M., Pang, A.T., and Lodha, S.K. (1996). Glyphs for Visualising Uncertainty in Vector Fields. *IEEE Transactions on Visualisation and Computer Graphics*, 2/3: 226–79.

Wolff, R.S. and Yaeger, L. (1993). *Visualisation of Natural Phenomena*. New York: Springer Verlag.

6

Interaction Design for Environmental Applications

6.1 Introduction

'Interaction' means that when one performs an action, there will be changes in what happens next (Robinett, 1994). In today's major computer user interfaces, one acts through an input device and can then see the results of that action on the computer screen. This type of interface, called direct manipulation (Shneiderman, 1988), requires the user to initiate all tasks explicitly and to overview all the events. Most environmental software is based on this user interface.

Developments on direct manipulation interface designs for environmental applications include the use of filters, sketching, sound, virtual environments, ubiquitous computing, and collaborative tools. These developments are reviewed in this chapter and illustrative examples provided.

As a cautionary note, it should be mentioned that many of these proposed alternative interactive modes have not been subject to formal usability testing. By usability testing (Nielsen, 1993), is meant obtaining measures on: error rates, time to complete tasks, learning, and workload; and qualifying subjective responses from the user. In virtual environments, an additional usability measure relates to cyber nausea.

Maes (1994) has argued that for non-technical users the direct manipulation metaphor is inappropriate. She proposes that a complementary style of interaction called indirect management should be adopted. In indirect management, the user is engaged in a cooperative process in which human and computer agents both initiate communication, monitor events, and perform tasks. Agents, in this context, represent personal assistants collaborating with the user. Although there are a very limited number of environmental applications using agents, this concept is potentially relevant for an area where predominantly non-technical individuals are required to participate in major studies.

6.2 Direct Manipulation

In previous chapters, methods for sensing, modelling, and visualising multi-dimensional data types have been reviewed. Software applications to perform these tasks and report their results are used in palmtops, notebooks, desktops, and shared screens.

Most of the available and foreseen environmental applications (standalone and networked) rely on interactive devices which include:

- Icons that represent objects such as files, folders, briefcases, and bins that can be opened by pointing and clicking, added, deleted, or dragged.
- Menus providing access to applications and documents by pointing and clicking.
- Sensitive text to pointing and clicking or to mouse rollovers that may lead to hyperlinks in hypertext applications.
- Props or objects that can be clicked on to provide information, such as about animals, pollution control devices, or buildings in an environmental application.
- Characters that have been used mainly in games for environmental education or as assistants in desktop applications, to guide users.
- Opening doors, collisions, and proximity of visibility operators that trigger actions when exploring an application.
- Structure as an interactive device, such as in Virtual Tejo (see Chapter 7), a three-dimensional quasi-realistic representation of a natural ecosystem. By exploring the representation, the user is obliged to learn and act on the virtual ecosystem.
- Garrand (1997) also refers to other examples, such as the exploration of buildings and cities, along the same lines.

In networked environments such as the Web, new interactive devices have appeared such as:

- Queries providing immediate feedback in the form of counters and graphs on user reactions to surveys. Many database backed Web sites provide these devices.
- Videoconferencing tools and chatrooms, as reviewed later in this chapter.

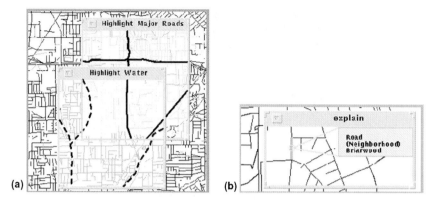

Fig. 6.1 Examples of Magic Lenses Filters for minimising display clutter and providing additional information: (a) highlighting map elements; (b) explaining map elements

Reprinted from Fishkin and Stone, 1995, CHI '94 © 1994 Association for Computing Machinery

6.3 Filters

Magic Lenses Filters (Fig. 6.1) proposed by Fishkin and Stone (1995) are relevant interactive devices for environmental applications. Magic Lenses use a lens emulator that changes the view of the objects viewed through that operator. By spatially bounding the filter, one can gain from the user's experience with physical lenses. A result already pointed out in Chapter 5 is the reduction in clutter in representations such as complex networks and diagrams (Fishkin and Stone, 1995).

Figure 6.1 illustrates how such filters can be used, in addition, to highlight information (Fig. 6.1a) and provide explanations on map elements (Fig. 6.1b).

Neves (1999) has proposed that filters inspired by animal vision may be used to browse spatial information in virtual environments. In Fig. 6.2, a filter is used to focus on a certain area while flying in a virtual world.

6.4 Sketching

Humans drew before they could write or even speak. Drawing is thus one of the most intuitive methods for human communication. Sketching is rapidly carried out drawing in order to satisfy some personal purpose.

Fernandes *et al.* (1997) proposed using sketching tools to interact with mosaics of orthophotos. These tools enable the interactive digitising of points, lines, and polygons over a background of digital orthophotos. Vector data

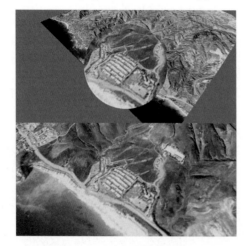

Fig. 6.2 Filters used to focus on a target area while flying in a virtual world

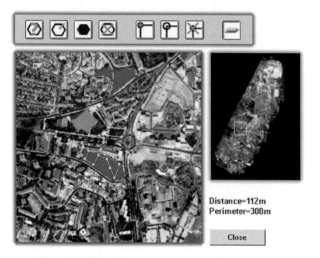

Fig. 6.3 Sketching on a mosaic of air photos

created with this procedure can then be used to define thematic layers (i.e., roads, streams, and administrative boundaries), associate these layers with information, and store them in a database. These vector data can be hyperlinked to any type of multimedia information (video, images, and sound) connected to a Universe Resource Locator (URL), creating a distributed spatial multimedia information system. This tool, implemented in Java, can also be used for spatial measurements such as area calculation, perimeters, line length, or distance between two points (Fig. 6.3).

Fig. 6.4 A general interface for browsing orthophotos, including fly-over, sketching, and dynamic sketching

Reprinted from *Computers & Geosciences*, Vol. 23, J.P. Fernandes et al., Visualization and Interaction Tools for Aerial Photograph Mosaics, pp. 465–74, Copyright 1997, with permission from Elsevier Science

A general interface to browse orthophotos is presented in Fig. 6.4. It includes icons for sketching, dynamic sketching (see Live Sketch in Chapter 4), fly-over (Chapter 5), and also change detection (Chapter 2).

The application of sketching tools can be greatly facilitated by the use of computer vision (Freeman *et al.*, 1998). Rather than sketching with a mouse or a pen on appropriate tablets (or expensive touch screens), users could use pointing commands and gestures to manipulate objects in the screen. This alternative is appropriate for shared screen applications, such as those that have been used or are being proposed for emergency planning and for environmental education applications. Current limitations are related to response times and reliability (Freeman *et al.*, 1998).

6.5 Sound

There are two types of sound relevant for user interface design: speech and non-speech sound.

Reddy (1976) proved that speech is four times faster than manual input and enables freedom of movement with the hands unoccupied. Today, there are products on the market with high degrees of accuracy and most work better if used by the same person all the time. Even so, there are always errors due to the ambiguity associated with language and variations in an individual's speech.

Shiffer (1993) has shown how geo-referenced voice annotation can be used for public participation in planning efforts. By looking at zoning proposals, for example, the user could record his/her comments concerning particular sites. Also, the browsing of a plan could include comments on specific proposals.

Buxton (1989) and Hereford and Winn (1994), discussing non-speech audio on the user interface, suggested mainly three alternative uses:

- alarms and warnings;
- status and monitoring indicators;
- encoded messages.

Patterson (1982) presented key concepts associated with the use of alarms and warnings, namely:

- Their objective is to warn the user by means of beeps and sirens.
- They should require minimal learning.
- They are based on digitised or synthesised sounds.
- They should attract the user attention in as unobtrusive a way as possible.

More recently, sound has been used in virtual environments to warn the user that he/she is moving out of range. In games, sound is often used to catch the user's attention (hearing commanding vision).

Gaver (1986) through his seminal work 'Sonic Finder' showed how one could create auditory icons to convey system status information. He recommended the use of everyday sounds to show, for instance, large messages hitting the mailbox or large files being dragged to the trash can, on the Macintosh interface. Auditory icons aimed to be caricatures of naturally occurring sounds requiring minimal learning. They rely on the principle that in the normal model of hearing, one listens to sounds in order to identify the events that cause them (Gaver, 1989). While auditory icons are based on the idea of using the dimensions of the sound's source (Gaver, 1989), earcons use the dimensions of sound to represent dimensions of data. Earcons have been proposed by Blattner *et al.* (1989) and have the following features:

- They convey information about objects, operations, and interactions.
- Musical elements are combined to convey meaning.
- Learning is required.
- Earcons use sampled sounds.

Silva and Silva (1995), Druin and Solomon (1996), and Silva *et al.* (1998) have reviewed commercial products for education that use sound in the user interface. A compelling environmental example is provided by Microsoft Oceans that tries to simulate the soundscapes of natural ecosystems and invites the matching of sounds to images of animals. A non-commercial example is provided by Silva *et al.* (1998) reviewed in Box 6.1

Box 6.1 *Catching Sounds in the Park*

Silva *et al.* (1998) proposed that in the synthetic representation of a natural ecosystem, children could use four sound tools to learn about the system (Fig. 6.5):

- *Headphones*: this tool may be used to discover and hear sounds by clicking on the images.
- *Information*: provides textual information about sounds and their sources.
- *Navigator*: using this tool and clicking on a sound, children will find a pull-down menu offering the possibility of travelling to sounds with similar ecological sources. Silva *et al.* (1998) refer to the example of clicking the sound of the bird chorus and then being able to travel to other bird sounds such as those from pigeons, geese, or blackbirds.
- *Catch a sound*: this tool is used to drag a sound to the right word (i.e., the sound of a goose to the honking word) or to the trash can if they find the sound in the wrong place (i.e., cow singing).

Figure 6.5 illustrates the interface of the proposed system that has been implemented for Parque Serralves in Porto, Portugal. The system was developed with the help of children who devised imaginative schemes for catching sounds as shown in Fig. 6.6.

Fig. 6.5 The interface for Catching Sounds in the Park. Sound tools include Headphones, Information, Navigator, and Catch a Sound

Reprinted from Silva *et al.* (1998) with permission from the Association for the Advancement of Computing in Education (AACE)

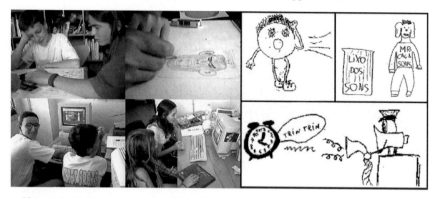

Fig. 6.6 Testing designs for Catching Sounds in the Park with children: the use of a trash can and a vacuum cleaner were among the ideas proposed

Reprinted from Silva *et al.* (1998) with permission from the Association for the Advancement of Computing in Education (AACE)

6.6 Virtual Environments

Virtual environments may be used to provide front-ends to the exploration of real spaces (both natural and artificial), information spaces, and communication spaces as illustrated by Barfield and Furness (1995). User interface design for communication spaces will be reviewed in Section 6.7. Design proposals for the interface with real and information spaces will be reviewed in this section.

6.6.1 Real Spaces

The most basic interaction mode in virtual environments is navigation (Camara *et al.*, 1998). There are now hundreds of virtual environments where one can navigate in natural and artificial spaces. Other interaction alternatives include object selection and manipulation, scaling, and the traditional menu and widget interaction. The Virtual Tejo, BITS, and Virtual GIS projects illustrate new concepts associated to the design of interfaces for environmental applications of virtual reality. Unexplored possibilities of tele-operators in real spaces are also proposed based on the idea of technological masquerading (Robinett, 1992).

Virtual Tejo

When designing a virtual reality interface for a real space, for example, an estuarine system, Camara *et al.* (1998) identified the following tasks:

- Fundamental tasks include problem perception, such as navigation (fly-over, walking, and diving), object manipulation (object selection, information, and retrieval), and visualisation (environmental quality levels).

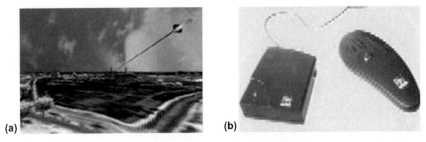

Fig. 6.7 (a) A virtual pointer in the Virtual Tejo user interface; (b) associated wire-less mouse

- Intermediate tasks involve navigation for problem investigation (terrain/object following) and visualisation (emission sources, compliance with standards).
- Advanced tasks are management actions, including the creation, deletion, scaling, and movement of natural and artificial entities.

Figure 6.7 shows an interface developed for the Virtual Tejo system which will be described in detail in Chapter 7. To perform the VR tasks a virtual pointer was selected. A wire-less mouse can control this pointer. To track the mouse position, a Polhemus Isotrack sensor was attached. The three mouse buttons had different functions:

- The central mouse button controlled object manipulation and was used to retrieve associated records from a database, modify its attributes (by changing the scale, position or behaviour), and create new objects or delete old ones (i.e., waste water treatment plants, buildings, trees).
- The right mouse button activates the navigation mode (terrain/object following or free exploration).
- The left mouse button triggers an environmental quality visualisation function. The virtual pointer is draped with red whenever the user navigates over a stretch of the water body violating legal standards.

BITS

Dias *et al.* (1995*a*,*b*) and Almada *et al.* (1996) developed a tool, named BITS (after Browsing In Time and Space), for a user to make annotations (logging) while exploring a virtual ecosystem. Earlier work in real-time annotation for video (Weber and Poon, 1994) and audio (Wittaker *et al.*, 1994) inspired the development of this tool.

BITS was designed following a virtual pad and pen metaphor (Fig. 6.8). It allows for travelling in space and time, navigation, and object selection and

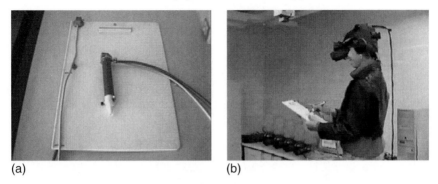

(a)   (b)

Fig. 6.8 BITS interface: (a) Virtual Notepad; (b) BITS in action

From Dias *et al.*, 1995*b*

manipulation, as well as logging. It is based on the principle of two-handed interaction proposed by Buxton and Meyers (1986) and the use of physical props in three-dimensional environments, such as those proposed by Stoakley *et al.* (1995).

Buxton and Meyers showed that users could easily transfer their everyday skills with two hands to the operation of a computer. Stoakley *et al.* (1995) proposed the World In Miniature (WIM), where a user has in his/her hands a miniature of the world he/she is exploring in a virtual environment. In WIM, an object can be manipulated either in miniature or in the original virtual environment, because its corresponding counterpart has the same modifications.

In BITS, the user is provided with a binaural view of the environment through the use of a head-mounted display. In the non-preferred hand, the user handles a notepad with an attached Polhemus sensor. The notepad incorporates a small joystick, with one degree of freedom that is manipulated by the user's forefinger. The notepad is where the user takes notes and issues commands. The other hand manipulates a pen that has an attached Polhemus sensor.

The virtual notepad is divided into six areas: Notes Area, Browse In Selection Buttons, Time and Space Slider, Pause Button, Log Button, and the Commands List. The Notes Area is the area where the user writes his/her notes and includes a scroll bar. The Browse In Selection Buttons is related to the use of the Time and Space Slider. This Slider facilitates travelling in space by selecting the Space Selection Button and travelling in time by selecting the corresponding button. Travelling in both time and space is done by selecting the Both Selection Button. The Pause Button enables the user to stop the simulation. To index the notes to a given step of simulation, the user presses the Log Button. Finally, the Commands List for the prototype shown in Fig. 6.9 includes six commands: Move, Bird, Fish, Open, Save, and Quit. The Move command is used to move objects during the simulation. The Bird and Fish commands automatically

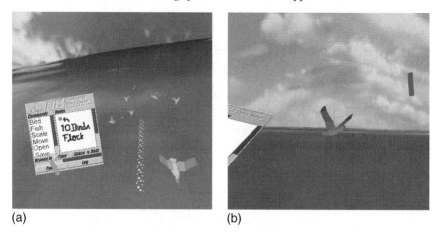

(a)   (b)

Fig. 6.9 BITS system in operation: (a) taking notes on the virtual notepad;
(b) picking up objects

From Dias *et al.*, 1995*b*

place the user near a bird or a fish, respectively. The Open command allows the user to load a notepad which was saved by a previous user.

Virtual GIS

According to Mark (1992), humans conceptualise and handle spatial data at three levels or spaces:

- *Haptic Space*, which is understood through haptic perception and sensori-motor experiences.
- *Pictorial Space*, which is mainly visually sensed.
- *Transceptual Space* is a mental production resulting from haptic and pictorial experiences over time. Wayfind metaphors in virtual environments are structuring paradigms for this space (Neves *et al.*, 1997).

Neves *et al.* (1997) suggested that by combining virtual environments and geographical information systems (GIS), these three spaces could be combined in one single computation environment called Virtual GIS Room, where:

- Users would work in the Haptic Space by using a digitising tablet, where a pen could be used to select parts of a map, paint the digital terrain model (DTM), select objects, and travel in virtual space (Fig. 6.10).
- Users could see different perspectives of data in the computer screen and invoke commands in the desktop environment. This Pictorial Space is the space typically used by GIS systems.
- Users can immerse themselves in the Transceptual Space, materialised by a head-mounted display, and explore the virtual space.

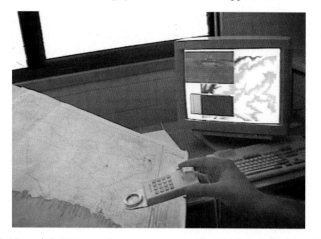

Fig. 6.10 Virtual GIS Room: the user is navigating in the virtual environment using the tablet mouse as a navigation interface

Reprinted from *Computers & Geosciences*, Vol. 23, J.N. Neves *et al.*, Cognitive Spaces and Metaphors: A Solution for Interacting with Spatial Data, pp. 483–8. Copyright 1997, with permission from Elsevier Science

Fig. 6.11 Virtual GIS Room: the workbench with a digital terrain model (DTM) can be draped with thematic maps (or images); these are posters on the room wall

The Virtual GIS Room is based on a workbench, where the DTMs are displayed in 2.5 dimensions (2D plus altimetry), with texture mapping of the images and several layers of information represented as 2D posters on the wall (Fig. 6.11). The users have direct access to the information in the posters and hear a 3D sound with a description of a layer after selecting it. The layers on the wall are 2D layers that change to 2.5D after draping the DTMs on the table (Neves *et al.*, 1997).

There have been several other efforts to bridge geographical information systems and virtual environments by Koller *et al.* (1995), Rhyne, and co-workers at

US EPA (1998). Koller *et al.* have developed an integrated real-time three-dimensional geographical information system. Rhyne *et al.* have converted information from ARC/INFO (GIS) to AVS (visualisation system) to VRML. They have also incorporated animation functions into digital elevation models available in VRML.

Technological Masquerading

Robinett (1992) suggests that new interfaces could be devised using tele-operation to study intraspecies communication in animals. He cites the example of a computer-controlled robot bee, which was able to direct real bees to specific locations far from the hive by performing the 'dance' that bees use to communicate. In addition, this robot bee also emitted a liquid sample of the distant pollen (Weiss, 1989).

6.6.2 Information Spaces

In the movie *Disclosure*, Michael Douglas is seen exploring the information systems of his company, represented as file cabinets, inside a virtual environment. A similar system was actually developed for Computer Associates by Sense8 in 1995. James Leftwich of Orbit Interaction has also designed a similar interface with a realistic representation of an office with doors, desks, and shelves (Anders, 1999).

Information Visualisation is the use of computer-supported interactive visual representations of abstract data to amplify cognition. Whereas scientific visualisation usually starts with a natural physical representation, Information Visualisation applies visual processing to abstract information. Card *et al.* (1991), Mackinlay *et al.* (1991), and Robertson *et al.* (1991) have proposed three-dimensional interfaces for information spaces as shown in Fig. 6.12. They are

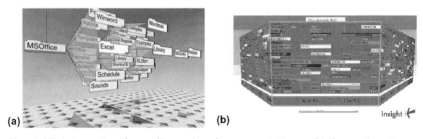

Fig. 6.12 Interactive three-dimensional representations of information structures for digital libraries: (a) Cone Trees; (b) Perspective Walls

Courtesy of the Xerox Corporation and Inxight Software, Inc.

computer-supported interactive visual representations of abstract data. Coors and Jung (1998) have recently proposed VRML interfaces to data warehouses using similar concepts.

6.7 Ubiquitous Computing

Ubiquitous computing was first proposed by Weiser (1991) by saying that 'the most profound technologies are those that disappear. They weave themselves in the fabric of everyday life until they are indistinguishable from it'. Ubiquitous computing has two main attributes (Weiser, 1991, 1993):

- *Ubiquity*, as interactions are not supposed to rely on a single workstation. Computation is accessible everywhere and wireless networks are widely available to support both remote and mobile access.
- *Transparency*, the technology is non-intrusive, invisible, and integrated in the human environment.

Norman (1998) and Abowd and Mynatt (2000) review the main developments and chart future research for ubiquitous computing. Many of their proposals are relevant for environmental engineering and sciences by supporting data collection tasks, project execution, and project team collaboration.

Data collection activities are benefiting from the array of mobile computing devices with telecommunications capabilities (and telecommunication devices with computing capabilities) that are already in the market. These devices usually incorporate global positioning system (GPS) receivers for positioning. However, these devices do not satisfy the 'transparency requirement'. The use of environmental sensors as part of wearable computers (Abowd *et al.*, 1997; Mann, 1997) provides a better illustration of the concept (see the Web site address in Section 6.10).

Lee *et al.* (2000) proposed HandScape, a measurement tape that is also a ubiquitous input device (see the Web site address in Section 6.10). The tape is complemented with orientation sensors. Thus, a user instead of just obtaining a linear distance captures a vector with both length and direction. A measuring tape becomes an input device for computer drawing and modelling.

To support project execution, two relevant proposals include MetaDesk and Digital Desk. MetaDesk developed by Ulmer and Ishii (1997) includes Tangible Geospace that uses physical models of landmarks, such as MIT's Great Dome and the Media Lab, as phicons (physical icons). These phicons allow users to manipulate two- and three-dimensional graphical maps of the MIT campus. Simultaneously, the arm-mounted 'active LENS' shows a three-dimensional view of the MIT campus. By grasping and moving the active LENS, the user can navigate the three-dimensional representation of the campus buildings.

Ubiquitous computing proposals for project executions include Wellner's Digital Desk. This 'desk' supports augmented interaction with physical paper documents on a physical desktop, identifying and augmenting these with overhead cameras and projectors (Wellner, 1993).

Ubiquitous computing proposals that may be applied to collaborative efforts are reviewed in the next section.

6.8 Collaborative Tools

Environmental projects demand collaborative tools as they are carried by teams (which often are multidisciplinary), and many studies (e.g., environmental impact statements, urban plans) require the public's participation. Computer based collaborative tools are listed in Table 6.1.

The most interesting tools from the standpoint of environmental applications and user interface design are:

- Shared screens, which may be important in disaster management, planning exercises, and environmental education.
- Videoconferencing systems enabling collaborative efforts involving team members remotely located.
- Chat systems for professionals and public participation.
- Groupware for project execution.
- Multi-user variants for environmental education.

Table 6.1 A taxonomy of collaborative tools

Same place–same time	*Same place–different time*
Shared screens	Interactive applications enabling annotation
Same time–different place	*Different time–different place*
Teleconference,	LISTSERV (mailing list programs for group
Audio conference	communication)
Videoconference	
	USENET (asynchronous text discussion on
Chat systems	several topics separated into newsgroups)
Multi-user variants	Groupware (collaboration systems that
(MUD, MUCK, MUSH,	include shared libraries, text, and graphic
MUSE, MOO)	environments and may be based on the WWW)

Note: MUD, multi-user domains; MUCK, multi-user collective kingdom; MUSH, multi-user shared hallucination; MUSE, multi-user shared environment; MOO, multi-user object oriented.

Source: Adapted from Wolz *et al.*, 1997.

This is a simplistic subdivision for clarity: these tools may overlap and can be used complementarily.

6.8.1 Shared Screens

Shared screens are well suited for project teams. They can replace the analogue boards with the digital advantages: storage of information, replay of historic information, and access to current information and simulations. Liveboard, proposed by Elrod *et al.* (1992), is a pioneer example. However, shared screens do present problems when large numbers of people want to interactively control the system, which for many functions, such as zooming or panning, is a technical impossibility. Robinett (1994) points out that such group interactions are limited to voting and elimination of entities (i.e., 'elimination' means 'shooting' in Robinett's article, as he was discussing the use of shared screens for entertainment).

In a technical setting, a shared screen can be divided into a number of shared screens if the system can be divided into as many subsystems. Each of the screens may have an associated projector such as Winograd's Interactive Works Spaces (Winograd, 1998). Each screen may also be a Liveboard as in Germany's National Research Centre for Information Technology (GMD) I-Land system (Fig. 6.13a,b), which is described in the Web site referred to in Section 6.10. Winograd and GMD's projects relate to the development of working murals using multiple projectors or Liveboards. Power Wall, developed at the University of Minnesota in cooperation with Silicon Graphics, is a high-end manifestation of the concept. Power Wall shows the potential of murals for high-resolution cooperative visualisations.

Winograd and GMD's systems also include horizontal shared screens in the form of working benches or tables (Fig. 6.13c). In a typical environmental impact assessment study, one can foresee meetings of groups of experts around working benches. In plenary meetings, those groups could project the work done on their benches on the mural thus making it available for discussion.

Hewlett Packard presented a video called Synergies at CHI'95, showing a crisis support infrastructure (for an earthquake occurring in the year 2000), that used ubiquitous computing and a shared screen (Hewlett-Packard, 1995). The infrastructure had headquarters in a room similar to a war room. Emergency planning units in many countries will certainly develop (if they have not already) a similar infrastructure as today's technology approaches the fiction of the past.

A high-end shared screen system is the Virtual Great Barrier Reef installation in the DOME immersive virtual reality system (Refsland *et al.*, 1999). Figure 6.14 displays the installation that was designed to hold 70–100 visitors interacting with 360 degrees over 360 degrees three-dimensional sights and sounds. The system was designed to make users feel involved in a story, not looking at a

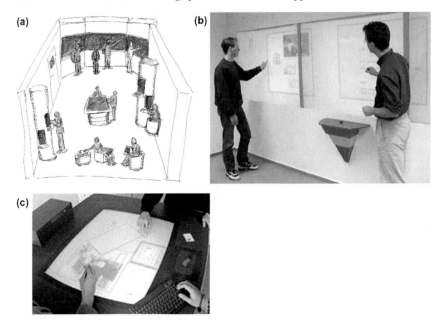

Fig. 6.13 GMD's I-Land System: (a) global vision; (b) shared screens use three
Liveboards; (c) working table

Courtesy of Norbert Streitz, GMD

Fig. 6.14 Virtual Great Barrier Reef installation

Courtesy of S. Thrane Refsland and Takeo Ojika

computer screen. An immersive 360 × 270-degree High Definition Television camera system, located under the water in the reef, transmits live pictures to the DOME. These images provide a background that is complemented by artificial life graphic fish. These use pattern recognition to identify the real fish of their own species and learn to emulate behaviours, collision detection, and depth perception (Refsland *et al.*, 1999). The environment includes also biobots, which are part human and part artificial life. Biobots enable users to play different marine life roles.

The DOME is made of an inflatable fabric that is highly reflective for projection viewing. From a computational standpoint, it is composed of a large distributed network, which uses four Power Walls (see above), and three CAVE systems (see Chapter 5). Interaction was done using computer vision techniques and input devices such as the Dive Panel and the FishGlove. The Dive Panel displays information on a small LCD monitor about the species under investigation and other information. The FishGlove was designed to facilitate control of the biobots.

6.8.2 Videoconferencing Systems

Videoconferencing systems have been widely used for many years, using over-the-air, microwave television signals or cable-transmitted compressed video (Schaphorst, 1996). They required expensive video and audio systems plus the cost of network or satellite television equipment.

With the Internet boom, small Web cameras and audio systems coupled to personal computers, videoconferencing is now affordable. With the projected increases in bandwidth, Internet videoconferencing will approach the quality of the traditional systems at a substantially lower cost. Videoconferencing fails to capture and is no substitute for many of the subtleties of human contact, however, improved design, software, and bandwidth may contribute improvements in the professional arena. Among available proposals in the research community, there are two that should be considered: ClearBoard (Ishii and Kobayashi, 1992; Ishii *et al.* 1994); and UbiMedia (Buxton, 1995).

ClearBoard enables the interaction, on the same screen, of users remotely located over a shared drawing as illustrated in Fig. 6.15. Both users can share a common orientation and can read all the text and graphics in their correct orientation. In addition, ClearBoard provides the capability that Ishii and co-workers call 'gaze awareness': the ability to monitor the direction of a partner's gaze and thus his/her focus of attention, something that cannot be achieved with a whiteboard.

UbiMedia intends to apply ubiquitous computing principles to videoconferencing. In traditional videoconferencing, the user is sitting in front of his/her computer talking to someone with a camera on top of his/her monitor. In

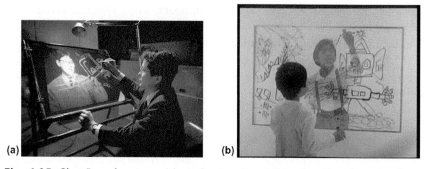

Fig. 6.15 ClearBoard system: (a) professional and (b) educational applications
Courtesy of Hiroshi Ishii and NTT

UbiMedia (a result of a system called UbiVivid), the user is free from the camera-focused interaction. This freedom is achieved by placing a large number of cameras and monitors throughout the environment. The video cameras are not seen as cameras, but as surrogate eyes, which means that the usual social conventions of human communication can be maintained. Two other relevant UbiMedia principles are (Buxton, 1995):

- The box into which solutions are designed is the room, not a box that sits on one's desk. This is a key difference between the ecological design of UbiMedia and the design of appliances.
- Every device used for interaction (cameras, microphones, loudspeakers) is a candidate for human–computer interaction and often simultaneously.

6.8.3 Chat Systems

On-line chat is one of the most (if not the most) popular uses of the Internet. Among the available chat facilities, Internet Relay Chat (IRC) dominates the field. IRC is international, multilingual, and its range of topics and freedom for corporate interests makes it widely attractive (Harris, 1995). To enter the world of IRC, a nickname is commonly adopted. The next step is the channel selection, which is done by first listing all the available channels and then joining the one of choice. Channels may be moderated. When one joins, one can see who the current participants of the channel are. At any time one can leave the channel and enter another channel, go away momentarily, or simply leave IRC altogether. It is always possible to hold in a buffer a history of one's conversation. When leaving, one can place a note to other users of IRC. IRC also allows the exchange of image and sound files.

Chat facilities, with simpler interfaces than IRC, are widely applied today in the Internet versions of newspapers, radio stations, and television networks. They are used to facilitate the interaction of personalities with the public and for consultation purposes on a wide range of areas including environmental problems.

They are already and will increasingly be, along with many other collaborative tools reviewed here, the method of choice for teams working with remotely located experts and for the public's collaboration and participation in environmental studies.

6.8.4 Groupware

Most groupware either relies or will rely on the current or future versions of the Internet. Research efforts involving environmental modelling are part of the National Computational Science Alliance, the United States high-performance computing initiative. In this initiative, the goal is to eliminate barriers of time and space separating collaborators across the United States by providing access to distributed data and high-resolution, high modality collaborative environments (Reed *et al.*, 1997). These will include CAVEs, ImmersaDesks (see Chapter 5) and Power Walls (see Section 6.8.1).

Reed *et al.* (1997), referring to previous studies, assert that recordings of results and histories are key components in collaborative data analysis. This means that collaboration environments should enable users to record interactions for future review when navigating in large datasets, something that can be accomplished with Virtual Director (Thiebaux, 1997). Another proposal by Reed *et al.* (1997) relates to the possibility of project team members to manipulate software components and their behaviour while immersed in shared virtual environments.

Rhyne (1998) has reviewed collaborative computing across the network for environmental applications. She indicated that in collaborative computing, resources become available to workers in a networked environment and this results in a metacomputer that could be used as easily as a personal computer. In a metacomputing environment, Web browsers are the tool of choice as they enable the running of animations and the exploration of three-dimensional worlds in VRML. Using Multicast Backbone (Mbone) further facilitates collaborations through the Internet. Mbone provides videoconferencing and enables the sharing of visual information (Macedonia and Brutzman, 1994). Rhyne also mentions that collaborative computing means that research problems may be spread across several computers (Rhyne, 1998), allowing computers to work in concert. A visualisation example is provided by LinkWinds (see Chapter 5).

There is other groupware for productivity purposes, Lotus Notes being a primary example. For information retrieval and organisation in the Internet, Vistabar is another relevant proposal (Marais and Bharat, 1997). Vistabar supports the authoring of annotations, the development of bookmark hierarchies, and the sharing of the findings by user communities.

6.8.5 Multi-user Variants

Multi-user domains, or MUDs, were originally role-playing games, largely text-based, similar to chatrooms but requiring taking an avatar (graphic representations that often bear no resemblance of the user) and moving within rooms. MUDs have evolved into graphic environments such as The Palace, Worlds, and AlphaWorld. They can take many forms such as MOOs (Multi-User Object Oriented), MUSHes (Multi-User Shared Hallucination), MUSEs (Multi-User Shared Environment), and MUCK (Multi-User Collective Kingdom). The MUD concept has been used in research units such as the Upper Atmospheric Research Collaboratory (UARC) (now SPARC) in Michigan and in the Collaboratory for Environmental and Molecular Sciences in the Pacific Northwest National Laboratory.

The collaboratory concept (Kouzes *et al.*, 1996) illustrates how researchers and professionals working remotely (see Fig. 6.16) may carry out the interaction with environmental data and models. The SPARC interface for access to data and models in the upper atmosphere (Figs 6.17 and 6.18) is an inspiration for designing environmental collaboratories.

MUDs enable the simulation of four key human activities (Robinett, 1994): look around; move through the world and see it from different viewpoints;

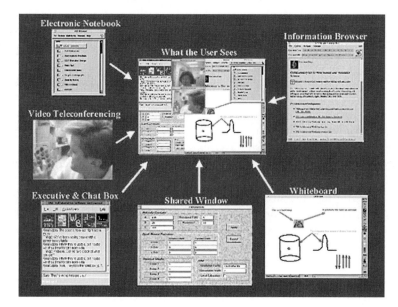

Fig. 6.16 Collaboratory capabilities: mail, videoconferencing, visualisations, whiteboard, and chatrooms

From Kouzes *et al.*, 1996 with permission from the IEEE Computer Society

Fig. 6.17 SPARC Instrument Status Map across the world

Source: http://www.crew.umich.edu/SPARC/index.htm

perform actions that can change the world; and talk with other people. It is no surprise that computer games have been based on the MUD concept. They follow several principles that include (Anders, 1999):

- Users can be represented by avatars.
- Avatars are controlled by keyboard, mouse, or other interactive devices.
- Avatars move through a spatial simulation, in many cases a 3D graphic environment.
- Action is sometimes seen through the eyes of the avatar.
- Players may encounter other avatars whose behaviour is driven by the gaming software.
- The simulated space of the game creates an arena for action and opportunity.
- Finally, players can often stop and save a game to resume it later.

These principles can be used to develop ecological games for environmental education. In fact, as part of a proposal for Expo98 (see Chapter 7), the author's research team suggested the exploration of coastal ecosystems in a virtual environment by a large group of users who could represent themselves, or terrestrial, and marine animals.

Multi-user simulations in virtual environments tend to be implemented as distributed interactive simulations (DIS). These were first proposed by the US Army's SIMNET. SIMNET took flight simulators and tank training

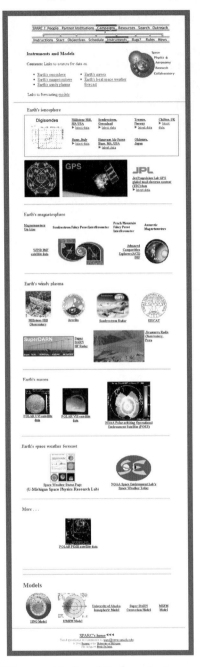

Fig. 6.18 SPARC's access to live data from instruments and models

Source: http://www.crew.umich.edu/SPARC/index.htm

simulators and networked them over telephone lines. Simulated battles involving over 1000 participants have been fought with this system (Robinett, 1994). The success of SIMNET and other DIS efforts were presented by Robinett as arguments in favour of creating interactive virtual worlds shared by a large number of users (Robinett, 1994).

Hoxie *et al.* (1998) have discussed standards for an Internet based DIS. They note that the DIS protocol consists of twenty-seven different kinds of messages, known as protocol data units (PDUs). These units are broadcasted over User Datagram Protocol/Internet Protocol (UDP/IP) networks. A PDU can describe an object's position or an event. DIS currently uses dead reckoning algorithms to predict the position of an entity. These predictive algorithms are used to minimise broadcasts with information on the state of the system entities. When the error of that prediction exceeds a certain threshold, the sending entity will update the network with its new state (Hoxie *et al.*, 1998).

6.9 Agents

Agents are 'soft' robots which assist users by being guides, memory aids, filters, matchmakers, educators, and entertainers (Miller, 1997). They are autonomous, act on behalf of a user, learn or adapt, and can migrate (virtually or physically). Entities that are sometimes erroneously considered as agents include (Miller, 1997):

- Software embodied in or serving an application, such as spell checkers in a word processor.
- Search engines, where there are explicit commands.
- Multimodal user interface systems, such as those using voice commands.

Maes (1994) evaluated the existing methods to develop interface agents according to two criteria:

- Competence, how does the agent acquire the knowledge it needs to decide when to help the user, what to help the user with and how to help the user?
- Trust, how can one guarantee that the user is comfortable when delegating tasks to an agent?

The available methods fall into three categories:

- The end-user programming approach, where the user creates rule based agents by endowing them with explicit knowledge. This method relies heavily on knowledge and effort from the end-user and, thus, scores low on the competence criterion (Maes, 1994).

- The knowledge-based approach, where the agent is provided with extensive domain-specific background knowledge about the application and the user. This method places high demands on the knowledge engineer and may lead to mistrust from the user, such as the agent's qualifications and sophistication.
- The machine learning approach, where the interface agent learns by observing the user's behaviour, from instructions he/she provides (this may use programming by demonstration, see Chapter 4) or from other more seasoned agents (Maes, 1994 and Lashkari *et al.*, 1998). This approach also scores well on the trust criterion because users can see how the agents' competence evolves.

There are very few examples of the use of interface agents in environmental (or even more generally, spatial) analysis. Campos *et al.* (1996) describe a knowledge-based interface agent for ARC/INFO that processes user's requests in natural language. The agent then translates this information into commands that ARC/INFO can understand. If they are not, the agent interacts with the user for clarification. If they are, the agent presents the results to the user.

Rodrigues and Raper (1999) have also proposed an interface agent architecture for the Smallworld geographical information system (GIS). This architecture includes an Agent Controller that monitors the GIS, and the Task Agents. Each Task Agent automates a specific GIS tool and/or helps the user learn how to use it. In the developed prototype, there was only one agent for the drawing and plotting tool.

6.10 Further Exploration

The following list of sites is periodically updated at (http://gasa.dcea.fct.unl.pt/camara/index.html). Student projects on user interface design for environmental applications can also be found at this site.

General Sites for Direct Manipulation Interfaces

The Special Interest Group on Computer Human Interaction (SIGCHI) of ACM has an excellent resource site (http://www.acm.org/sigchi).

Approximately 20,000 bibliographical and resource records on human–computer interaction may be found at (http://www.hcibib.org).

Ben Shneiderman's Web site supplements his book (*Designing the User Interface*, 1998), with course aids and a list of Web resources (http://aw.com/DTUI).

Usability testing is covered at Jakob Nielsen's Web site (http://www.useit.com). A comprehensive treatment of usability testing applied to Web design may be found at (http://usableweb.com).

Magic Lenses

Ken Fishkin's Web site includes most papers related to the use of Magic Lenses (http://www.parc.xerox.com/istl/members/fishkin/).

Sketching

The work by Fernandes *et al.* (1997) and Nobre and Camara (1999) related to the use of sketching may be found at (http://virtual.dcea.fct.unl.pt/~jmr/sem97/index.htm).

Sound

The work by Silva *et al.* (1998) on catching sounds is available at (http://gasa.dcea.fct.unl.pt/gasa/gasa98.htm). A leading research group on the use of audio is MIT Media Lab Speech Group (http://www.media.mit.edu/speech). At this site, see the Dynamic Soundscapes project that enables audio browsing in time and space.

Virtual Environments

Chapter 5 includes a comprehensive list of sites on virtual reality interfaces.

Information Spaces

Inxight is the leading firm in the development of information spaces (http://www.inxight.com).

Ubiquitous Computing

Mark Weiser's classic Web site on ubiquitous computing is at (http://www.ubiq.com/hypertext/weiser/UbiHome.html). Donald Norman's The Invisible Computer book Web site is (http://www.jnd.org).

Hiroshi Ishii's research on Tangible Bits may be found at (http://www.media.mit.edu/~ishii).

Wearable computing is covered at Steve Mann's site (http://wearcam.org/mann.html).

Links to sites covering Internet connections to robots, Web cameras, and other data acquisition devices are included in Yahoo's (http://dir.yahoo.com/Computers_and_Internet/Internet/Interesting_Devices_Connected_to_the_Net/). Bill Buxton maintains a review of input devices (http://www.billbuxton.com).

Ricochet, a wire-less Internet provider via radio, is located at (http://ricochet.net). For information on Internet access on mobile phones see (http://www.waptorum.org) and (http://gelon.net).

Shared Screens

Terry Winograd's group research on Interactive Work Spaces is described at (http://graphics.stanford.edu/projects/iwork). The site has links to major international projects on the use of murals and responsive workbenches. Two examples are University of Minnesota's Power Wall (http://www.lcse. umn.edu/research/powerwall/powerwall.html) and the Darmstadt I-Land project (http://www.darmstadt.gmd.de/ambiente/i-land.html).

Conferencing Systems

Conferencing systems available in the Internet are numerous. Two illustrative examples are San Francisco State University's COW system (http://thecity.sfsu.edu/COW2/) and CU–See-Me (http://cuseeme.com/), originally developed at Cornell University.

ICQ (I Seek You) enables synchronous and asynchronous conferencing (http://www.mirabilis.com/).

Papers on ClearBoard and related systems by Hiroshi Ishii are available at his site (http://www.media.mit.edu/~ishii).

Chat Systems

The most popular chat system is MIRC (http://www.mirc.co.uk/index.html). NetMeeting with audio and video capabilities has been used for professional and teaching purposes (http://microsoft.com).

Groupware

The University of Georgia maintains a generic site on groupware (http://www.terry.uga.edu/groupware). Lotus Notes developments are available at (http://www.lotus.com).

Multi-user Variants

The Contact Consortium (http://www.ccon.org) includes links for forums such as Worlds (http://www.worlds.com). The Consortium is a resource site on virtual communities and virtual worlds on the Internet.

Sites on collaboratories include:

- Northwestern Visualization Project for the geosciences (http://www.covis.nwu.edu).
- Collaboratory for Environmental Molecular Sciences (http://www.emsl.pnl.gov:2080/docs/collab/).
- Space Physics and Aeronomy Research Collaboratory (ex-UARC) (http://www.crew.umich.edu/SPARC/index.htm).

An illustrative example of a bulletin board is provided at (http://www.infopop.com). Twiki is a state-of-the art collaboration tool (http://twiki.sourceforce.net).

Mak is a leading firm in distributed interactive simulation (DIS) (http://www.mak.com). DIVE (Distributed Interactive Virtual Environments) is an Internet based multi-user virtual reality system where participants navigate in three-dimensional space and interact with other users and applications (http://www.sics.se/dive/dive).

Agents

The MIT Media Lab Agents Group Web site is at (http://agents.www.media.mit.edu/groups/agents/). A compilation of links on agents may be found at (http://www.cs.bham.ac.uk/~amw/agents/links/index.html).

References

Abowd, G.D. and Mynatt, E.D. (2000). Charting Past, Present and Future Research in Ubiquitous Computing. *ACM Transactions on Computer-Human Interaction*, 7/1: 29–58.

Abowd, G., Dey, A., Orr, R., and Brotherton, J. (1997). Context-Awareness in Wearable and Ubiquitous Computing. In *Proceedings of IEEE 1st International Symposium on Wearable Computers*. Cambridge, MA: IEEE Press. Available at http://www.cs.gatech.edu/fce/pubs/iswc97/wear-poster.html.

Almada, A., Dias, A.E., Silva, J.P., Pedrosa, P., and Camara, A.S. (1996). Exploring Virtual Ecosystems. *Proceedings of ACM AVI '96 Advanced Visual Interfaces*, 166–74. Gubbio, Italy: ACM Press.

Anders, P. (1999). *Envisioning Cyberspace, Designing 3D Electronic Spaces.* New York: McGraw-Hill.

Barfield, W. and Furness, T.A. (1995). *Virtual Environments and Advanced Interface Design.* New York: Oxford University Press.

Blattner, M.M., Sumikawa, D.A., and Greenberg, R.M. (1989). Earcons and Icons: Their Structure and Common Design Principles. *Human–Computer Interaction*, 4/1: 11–44.

Buxton, W. and Myers, B.A. (1986). A Study in Two-Handed Input. *Proceedings of ACM CHI '86*, 321–6. New York: ACM Press.

Buxton, W. (1989). Introduction to This Special Issue on Nonspeech Audio. *Human–Computer Interaction*, 4/1: 1–9.

Buxton, W. (1995). Ubiquitous Video. *Nikkei Electronics*, 3/27: 181–95.

Camara *et al.* (1998). Virtual Environments and Water Quality Management. *Journal of Infrastructure Systems*, ASCE, 4/1: 28–36.

Campos, D., Naumov, A.Y., and Shapiro, S.C. (1996). Building an Interface Agent for ARC/INFO. *ESRI User Conference Proceedings*, Palm Springs, CA.

Card, S.K., Robertson, G.G., and Mackinlay, J.D. (1991). The Information Visualise: An Information Workspace. *Proceedings of ACM CHI '91*, 181–8. New York: ACM Press.

Coors, V. and Jung, V. (1998). Using VRML as an Interface to the 3D Data Warehouse. *Proceedings of VRML98*, 121–7. Monterey Bay, CA: ACM Press.

Dias, A.E., Silva, J.P., and Camara, A.S. (1995*a*). BITS: Browsing in Time and Space. *Companion Volume ACM CHI '95*, 248–9. Denver, CO: ACM Press.

Dias, A.E., Almada, A., Silva, J.P., Pedrosa, P., Santos, E., and Camara, A.S. (1995*b*). Exploring Virtual Ecosystems with BITS. *Proceedings of 1st Conference on Spatial Multimedia and Virtual Reality*, Lisbon, Portugal.

Druin, A. and Solomon, C. (1996). *Designing Multimedia Environments for Children: Computers, Creativity, and Kids*. New York: Wiley.

Elrod, S., Bruce, R., Gold, R., Goldberg, D., Halasz, F., Janssen, W., *et al.* (1992). LIVE-BOARD: A Large Interactive Display Supporting Group Meetings, Presentations, and Remote Collaboration. *Proceedings of ACM CHI '92*, 599–607. Monterey, CA: ACM Press.

Fernandes, J.P., Fonseca, A., Pereira, L., Faria, A., Figueira, H., Henriques, L., *et al.* (1997). Visualisation and Interaction Tools for Aerial Photograph Mosaics. *Computers & Geosciences*, 23/4: 465–74.

Fishkin, K. and Stone, M.C. (1995). Enhanced Dynamic Queries via Movable Filters. *Proceedings of ACM CHI '95*, 8–11. Denver, CO: ACM Press.

Freeman *et al.* (1998). Computer Vision for Interactive Computer Graphics. *IEEE Computer Graphics and Applications*, 18/3: 42–53.

Garrand, T. (1997). *Writing for Multimedia*. Boston: Focal Press.

Gaver, W.W. (1986). Auditory Icons: Using Sound in Computer Interfaces. *Human–Computer Interaction*, 2/2: 167–77.

Gaver, W.W. (1989). The SonicFinder: An Interface That Uses Auditory Icons. *Human–Computer Interaction*, 4/1: 67–94.

Harris, S. (1995). *The IRC Survival Guide*. Reading, MA: Addison-Wesley.

Hereford, J. and Winn, W. (1994). Non-Speech Sound in Human-Computer Interaction: A Review and Design Guidelines. *Journal of Educational Computing Research*, 11/3: 211–33.

Hewlett-Packard (1995). *Synergies*. Video presented at CHI '95. Denver, CO: ACM Press.

Hoxie, S., Irizarry, G., Lubetsky, B., and Wetzel, D. (1998). Developments in Standards for Networked Virtual Reality. *IEEE Computer Graphics and Applications*, 18/2: 6–9.

Ishii, H. and Kobayashi, M. (1992). ClearBoard: A Seamless Media for Shared Drawing and Conversation with Eye Contact. *Proceedings of ACM CHI '92*, 525–32. Monterey, CA: ACM Press.

Ishii, H., Kobayashi, M., and Arita, K. (1994). Iterative Design of Seamless Collaboration Media. *Communications of the ACM* (Special Issue on Internet Technology), 37/8: 83–97.

Ishii, H. and Ullmer, B. (1997). Tangible Bits: Towards Seamless Interfaces between People, Bits and Atoms. *Proceedings of ACM CHI '97*, 234–41. Atlanta, GA: ACM Press.

Koller, D., Lindstrom, P., Ribarsky, W., Hodges, L.F., Faust, N., and Turner, G. (1995). *Virtual GIS: a Real-time 3D Interface for Geographical Information Systems*. Technical

Report, Atlanta: Georgia Tech. Available at ftp://ftp.gvu.gatech.edu/pub/gvu/tech-reports/95–14.ps.Z

Kouzes, R.T., Myers, J.D., and Wulf, W.A. (1996). Collaboratories: Doing Science on the Internet. *IEEE Computer*, 29/8: 40–6.

Lashkari, Y., Metral, M., and Maes, P. (1998). Collaborative Interface Agents. In M. Huhns and M. Singh (eds.), *Readings in Agents*. San Francisco: Morgan Kaufmann.

Lee. J., Su, V., Ren, S., and Ishii, H. (2000). Handscape: A Vectorizing Tape Measure for On-Site Measuring Applications. *Proceedings of CHI '2000*, 137–44. New York: ACM Press.

Macedonia, M.R. and Brutzman, D.P. (1994). Mbone Provides Audio and Video Across the Internet. *IEEE Computer*, 27/4: 30–6.

Mackinlay, J.D., Robertson, G.G., and Card, S.K. (1991). The Perspective Wall: Detail and Context Smoothly Integrated. *Proceedings of ACM CHI '91*, 173–9. New York: ACM Press.

Maes, P. (1994). Agents that Reduce Work and Information Overload. *Communications of the ACM*, 37/7: 31–40.

Mann, S. (1997). Wearable Computing: A First Step Toward Personal Imaging. *IEEE Computer Magazine*, 30/2: 25–32.

Marais, H. and Bharat, K. (1997). Supporting Co-operative and Personal Surfing with a Desktop Assistant. *Proceedings of ACM UIST '97*, 129–38. Banff, Alberta: ACM Press.

Mark, D. (1992). Spatial Metaphors for Human-Computer Interaction. *Proceedings of 5th International Symposium on Spatial Data Handling*, Charleston, SC.

Miller, M. (1997). Software Agents. CHI 97 Tutorial, Los Angeles, CA.

Neves *et al.* (1997). Cognitive Spaces and Metaphors: A Solution for Interacting with Spatial Data. *Computers & Geosciences*, 23/4: 483–8.

Neves, J.N. (1999). Filters for Browsing Virtual Environments. Unpublished paper, Imersiva, Monte de Caparica, Portugal.

Nielsen, J. (1993). *Usability Engineering*. Boston, MA: Academic Press.

Nobre, E. and Camara, A. (1999). Spatial Simulation by Sketching. In A.S. Camara and J. Raper (eds.), *Spatial Multimedia and Virtual Reality*, 103–10. London: Taylor & Francis.

Norman, D.A. (1998). *The Invisible Computer*. Cambridge, MA: MIT Press.

Patterson, R.R. (1982). *Guidelines for Auditory Warning Systems on Civil Aircraft*, Paper No. 82017. London: Civil Aviation Authority.

Reddy, D.R. (1976). Speech Recognition by Machine: A Review. In N.R. Dixon and T.B. Martin (eds.), *Automatic Speech and Speaker Recognition*, 56–86. New York: IEEE Press.

Reed, D.A., Giles, R.C., and Catlett, C.E. (1997). Distributed Data and Immersive Collaboration. *Communications of the ACM*, 40/11: 39–48.

Refsland, S., Ojika, T., Loeffler, C., and DeFanti, T. (1999). Virtual Great Barrier Reef. In J.C. Heudin (ed.), *Virtual Worlds*, 153–79. Reading, MA: Perseus.

Rhyne, T., Brutzman, D., and Macedonia, M. (1997). Internetworked Graphics and the Web. *IEEE Computer*, 30/8: 99–101.

Robertson, G.G., Mackinlay, J.D., and Card, S.K. (1991). Cone Trees: Animated 3D Visualisations of Hierarchical Information. *Proceedings of ACM CHI '91*, 189–94. New York: ACM Press.

Robinett, W. (1992). Synthetic Experience: Proposed Taxonomy. *Presence*, 1/2: 229–47.

Robinett, W. (1994). Interactivity and Individual Viewpoint in Shared Virtual Worlds: The Big Screen vs. Networked Personal Displays. *Computer Graphics*, 28/2: 127–30.

Rodrigues, A. and Raper, J. (1999). Defining Spatial Agents. In A.S. Camara and J. Raper (eds.), *Spatial Multimedia and Virtual Reality*. London: Taylor & Francis.

Schaphorst, R. (1996). *Videoconferencing and Videotelephony: Technology and Standards*. Boston, MA: Artech House.

Shiffer, M. (1993). Augmenting Geographic Information with Collaborative Multimedia Technologies. *Proceedings of AUTOCARTO 11*, 367–76, Minneapolis, MN.

Shneiderman, B. (1988). Direct Manipulation: A Step Beyond Programming Languages. *IEEE Computer*, 16/8: 57–69.

Shneiderman, B. (1998). *Designing the User Interface: Strategies for Effective Human-Computer Interaction*. Reading, MA: Addison-Wesley.

Silva, M.J. and Silva, J.P. (1995). Looking and Hearing Glasses: Travelling Tools in Environmental Education. *Proceedings 1st Conference on Spatial Multimedia and Virtual Reality*, 181–6. Lisbon: New University of Lisbon.

Silva, M.J., Silva, J.P., and Hipolito, J. (1998). They Are Catching Sounds in the Park! Exploring Multimedia Soundscapes in Environmental Education. *Proceedings of Ed-Media & Ed-Telecom 98*, 1318–24. Charlottesville, VA: Association for the Advancement of Computing in Education.

Stoakley, R., Conway, M., and Pausch, R. (1995). Virtual Reality on a WIM: Interactive Worlds in Miniature. *Proceedings of ACM CHI '95*, 265–72. Denver, CO: ACM Press.

Thiebaux, M. (1997). Virtual Director. Master's Thesis, Department of Electrical Engineering and Computer Science, University of Illinois at Chicago.

Ullmer, B. and Ishii, H. (1997). The metaDESK: Models and Prototypes for Tangible User Interfaces. *Proceedings of ACM UIST'97*, 223–32. Banff, Alberta: ACM Press.

Weber, K. and Poon, A. (1994). Marquee: A Tool for Real-Time Video Logging. *Proceedings of CHI '94*, 58–64. Boston, MA: ACM Press.

Weiser, M. (1991). The Computer for the 21st Century. *Scientific American*, 256/3: 66–75.

Weiser, M. (1993). Some Computer Science Issues in Ubiquitous Computing. *Communications of the ACM*, 36/7: 75–84.

Weiss, R. (1989). New Dancer in the Hive. *Science News*, 136/8: 282–3.

Wellner, P. (1993). Interacting with Paper on the Digital Desk. *Communications of the ACM*, 36/7: 86–96.

Winograd, T. (1998). Interactive Work Spaces—A Human Centered Architecture. Unpublished paper, Stanford University, Palo Alto, CA. Available at http://graphics.stanford.edu/projects/iwork.

Wittaker, S., Hyland, P., and Wiley, M. (1994). Filochat: Handwritten Notes Provide Access to Recorded Conversations. *Proceedings of CHI'94*. Boston: ACM Press.

Wolz, U., Palme, J., Anderson, P., Chen, Z., Dunne, J., Karlsson, G., *et al.* (1997). Computer-Mediated Communication in Collaborative Educational Settings. *Report of the ITiCSE 97, Working Group on CMC in Collaborative Educational Settings*.

7

Towards Interactive Environmental Movies

7.1 Interactive Movies

Korolenko (1997) maintains there is no such thing as an interactive movie. No one is interested in sitting in a theatre and pressing a button whenever there is a menu of choices. Traditional movies are, for Korolenko, a much more attractive proposition. However, Altman and Nakatsu (1997), in their introduction to interactive movies, lead one to believe that there may be successful interactive movies on environmental problems. Interactive movies, in Altman and Nakatsu terms, are viewed as being interactive stories, interactive experiences between actors and audience and/or as having interactive movie actors.

Traditional movies rely on techniques that can be the basis for compelling interactive storytelling. They include (adapted from Altman and Nakatsu, 1997):

- *Non-linear story trees.* Storytelling may use maze, character, and alternative reality models. In the maze model, the story may be represented as a complex graph, where nodes are events. The character model means that the story is based on following a character. Finally, in the alternate reality model, there are multiple concurrent realities. Movies based on non-linear story trees may become interactive by introducing decision points. They can be implemented digitally using scripting languages.
- *Gradual discourse.* Many movies reveal their story gradually. Computerised interactive cinema influenced by this technique may divide the movie into levels of complexity. The viewer can then move along the different levels by pressing instruction icons.

- *Parallel action.* In this technique, the movie jumps back and forth to two concurrent events. In a digital implementation, the behaviour of characters have to be coordinated. The viewer may then command when to move from one event to the other.
- *Camera movement.* Techniques of camera movement may include the selection of viewpoint, focus, and level of detail. They can become interactive by using virtual reality technologies. A virtual reality implementation would also enable viewing movie scenes from different viewpoints as shown in Fig. 5.33.

These examples refer to the use of interactive techniques contributing to the enhancement of the traditional storytelling experience. However, digital techniques, such as virtual reality and artificial life models, are offering alternative movie experiences through the interaction between the audience and the actors. The Virtual Great Barrier Reef (see Chapter 6) is an illustrative example of an experiment between an audience and interactive actors using a shared screen. Artificial models running on virtual environments or video based simulations (Chapter 4) are techniques for creating interactive movie actors.

Testing interactive movie concepts was the goal of two projects led by the author: Environment Expo98 and Virtual Tejo. They are not interactive movies in a strict sense but they include some of their relevant features. Environment Expo98, an interactive application on Expo98 environmental studies, was a laboratory for non-linear story trees (Fonseca *et al.*, 1995*a,b*, Fonseca and Camara, 1997; Fonseca, 1998; Fonseca *et al.*, 1999). In Virtual Tejo, a water quality management system with a virtual reality front-end, camera movement techniques, and simulation models were tried out as a basis for interactive movies (Camara *et al.*, 1998). Interactive environmental movies, particularly those based in virtual reality, may be the opposite of Woody Allen's *The Purple Rose of Cairo*, where an actor leaves the movie to join real life. In those movies, one may leave reality and enter in the movie as an actor.

The main hypothesis underlying these projects was that interactive environmental movies could be the most appropriate channel to communicate with the media, public, and decision-makers. They can also complement written reports, numerical models, and geographical information systems to assist environmental professionals and scientists.

The research involved for Environment Expo98 and Virtual Tejo also provided a base for bringing together many of the concepts described in previous chapters.

7.2 Environment Expo98

Expo98 was held on a site that was totally renovated during 1993–8. The intervention area associated with the project included: an oil industry complex cover-

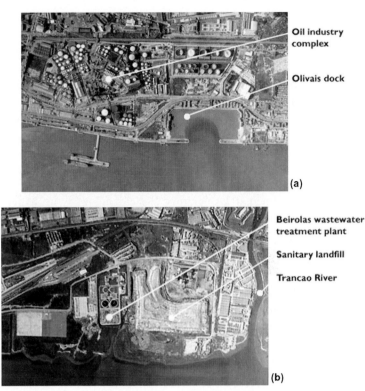

Oil industry complex

Olivais dock

(a)

Beirolas wastewater treatment plant

Sanitary landfill

Trancao River

(b)

Fig. 7.1 Aerial photos of the Expo98 intervention site showing environmental problems in 1993: (a) the western area; (b) the eastern area

Image courtesy of Parque Expo98

ing 50 hectares; Lisbon's slaughterhouse; an Army depot; a landfill and solid waste treatment plant; several harbour facilities; a waste water treatment plant; and it had a highly polluted river in the vicinity. Figure 7.1 highlights the major problems in 1993.

Data collected at the Expo98 site included all relevant parameters for water, air, and soil considering the land uses that were in place in 1993. In addition, meteorological, energy related, and fauna and flora data were also sampled. The firms in charge of the individual studies specified the data monitoring programmes. After 1998, the data monitoring programme continues. These data have been stored in the Expo98 Environmental Information System presented in Chapter 3.

This environmental information system provided the foundations for an experiment in designing a system that could support decisions and at the same time could be used to communicate with the media and the public—it was named the Expo98 Environmental Exploratory System. Four designs were made before it was decided that a Web enabled CD-ROM should be used. This system was technical in nature but issues normally associated with the production of multimedia titles for education and entertainment were present. It was also a tool for researching design of spatial multimedia hypermedia systems (Fonseca, 1998).

A document-centred approach was avoided from the start as rigid sequences and constraints are used to guide users in the navigational process. Instead, an exploratory approach was followed, where the user can explore the hypermedia system freely.

7.2.1 Structure

The Expo98 Environmental Exploratory System followed the hypermedia model in all four prototypes. This model was selected following Shneiderman's (1989) recommendations:

- The information on the Expo98 environmental projects was organised in fragments.
- The fragments related to each other.
- The potential user of the Exploratory System only needed a small fraction of the fragments at any one time.

The hypermedia model is based on the use of nodes, links and anchors, which define a graph. This graph provides the overall structure of a hypermedia system.

Nodes are pieces of information that may be a text, image, or video. They should convey only one idea or theme. Links are implemented if words, sentences, or images within a node can be associated with nodes containing related information. The starting point of a link is known as its anchor. Links may be divided into (Ginige *et al.*, 1995):

- *Structural links.* These are the key links that highlight the base structure of the system.
- *Associative links* that connect concepts that are related and, thus, may provide multiple views (or alternate realities) on a given piece of information.
- *Referential links* that provide additional information on terms used in the document.

There were three types of structural links in the Exploratory System: spatial, temporal, and thematic. They enabled the user to investigate the document in these three dimensions. The spatial links were set up from maps, aerial photos, and three-dimensional walkthrough models. By pressing buttons located on these images or using zooming facilities, one could move along regional and local dimensions, for each temporal stage. Four main temporal stages were considered: baseline conditions (1993), construction (1996), exhibition (1998), and post-exhibition (2000). To access spatial and thematic data for each of these temporal steps, a time slider was designed for the three last prototypes. Finally, the thematic links were divided into three major groups: those related to physical and chemical characteristics of the environmental components (i.e., water and air quality); those concerning biological features (i.e., fauna and flora); and those regarding cultural characteristics (i.e., aesthetics, recreation, man-made facilities and activities). One could explore these links, setting up, a priori, the spatial and temporal references.

Associative links (or multiple representations) were established within each group of structural links. For instance, when observing a certain area on an aerial photo, it was always possible to look at the map representation, ground photos and, in certain instances, to videos. When exploring a theme, one can also use associated visualisation and simulation tools.

Referential links were used to connect all technical terms used in the Exploratory System to a glossary.

7.2.2 Data Content and Tools

The types of data used in all Expo98 Environmental Exploratory System prototypes included:

- *Maps and air photos* (both vertical and oblique) on the whole area.
- *Photographs* of ground scenes.
- *Digital video.* Video was used to show backgrounds and point scenes, adding realism and providing a dynamic view of the area. At some key corner points, 360-degree circular views were used (i.e., around the Olivais dock in 1993 the images were dominated by the refineries; that view changed into the new buildings by 1998).
- *Synthetic three-dimensional images.* Computer aided design images on the proposed buildings, parks, and infrastructures for Expo98 were used.
- *Animated sequences.* Keyframe animation was used to show the evolution from 1993 to 1998 in certain areas (i.e., change of refinery tanks into buildings, change detected on aerial photos) and on the environmental quality data using scientific visualisation methods.
- *Audio.* Music was used in the first prototype. Experiences were also done with the representation of noise and with voice annotation methods. Indication of noise levels and pre-recorded noise sounds were provided when clicking in hot spots in the map of the area. System users could also insert oral comments by activating a voice icon and placing it on the area of their choice. In addition, oral explanations were used in association with videos included in the system.
- *Text*, with the reports on all major environmentally related projects.

There were three major tools to enable the exploration of these data types: Visualiser, Hyperbook, and Environmental Glossary. The Visualiser allows the direct access to animations, videos and photos organised in thematic catalogues. The Hyperbook is an electronic book that organises all the technical environmental reports developed during the Expo98 implementation phase as hypermedia documents. The available reports include the Environmental Impact Assessment Report and the Environmental Management Program. Finally, the Environmental Glossary enables users to search for definitions on more than 600 environmentally related terms. Some of these definitions are enhanced with videos and photos. The Glossary also includes several constants and basic functions in environmental studies.

7.2.3 User Interface Design

To develop the Expo98 Environmental Exploratory System, an iterative prototype–user testing approach was followed. There were three prototypes, before the final product. The book's Web site includes videos synthesising the first three prototypes. Storyboards were developed for each of the prototypes.

User interface design followed basic rules proposed by, among others, Anderson and Veljikov (1990):

- written text was kept simple and the number of words per page limited;
- processes were kept visual as far as possible;
- screen formats were consistently used;
- feedback and help facilities were provided to the users;
- navigation included linking (structural, associative, referential), next page/previous page, return to the main menu, and help;
- there was an attempt to keep users engaged.

Testing included concept testing within the team, review by peers and informal usability testing. Acceptance testing was performed only in the last prototype.

The first prototype (Fig. 7.2) was developed in 1994 based on the concept of providing multiple views of different key spots in the Expo98 intervention area. It included already demonstrations of a variety of tools to visualise, simulate, and understand the site and its environmental problems. The most attractive at the time was a morphing facility that enabled one to animate the expected transformations (from 1993 to 1998) in areas such as the refineries and the landfill. This prototype tried to bring traditional Environmental Impact Assessment (EIA) studies into a multimedia platform. However, the structure was inadequate and hampered navigation. It was also thought not to be sufficiently compelling for communication with the public.

The second prototype proposed in 1995 was, in fact, highly compelling (Fig. 7.3). A transparent air bubble was created, from where the user could direct

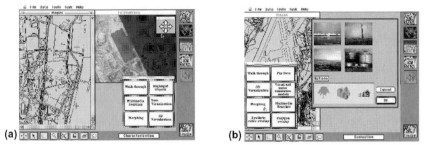

Fig. 7.2 User interface of the first prototype for the Expo98 Environmental Exploratory System. The emphasis was on providing multiple views: (a) maps and aerial photos; (b) maps and ground photos

Source: Fonseca, 1998. Video available on the book's Web site

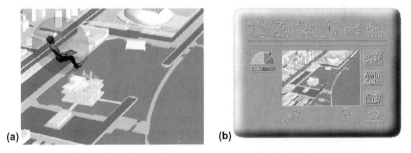

Fig. 7.3 User interface of the second prototype for the Expo98 Environmental Exploratory System. The navigating vehicle (a) became the focus of the interface (b)

Source: Fonseca *et al.*, 1995*b*. Video available on the book's Web site

his/her actions. As in a real vehicle, the air bubble had a control panel that was always accessible to the user. This panel would open up in the upper portion of the screen, a working area. However, this division of the screen into two sections greatly reduced the size of the working area and its legibility.

This fundamental criticism provided the basis for the design of the third prototype (Fig. 7.4), which was ready in 1997. The main screen included a large working area surrounded by icons with attached sounds. These icons enabled the access to the main structural, associative, and referential links, but number of icons was judged to be too high by users.

These three prototypes were implemented using a hypermedia–authoring tool for the Macintosh. The need for a version available across platforms led to the development of the final product as a Web enabling CD-ROM (Figs 7.5–7.8). For this version, all the visualisation and exploration capabilities were

Fig. 7.4 Control panel of the third prototype for the Expo98 Environmental
Exploratory System

Video available on the book's Web site

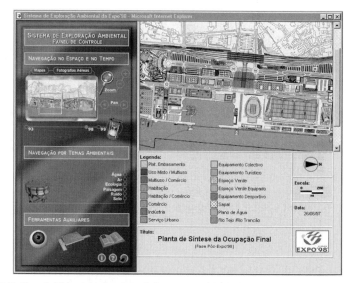

Fig. 7.5 Spatial Navigation Module: zooming and panning over the Expo98
site map

Source: http://sig.cnig.pt/websea

programmed with Dynamic HTML in association with JavaScript. This option
enabled keeping the free exploratory features of the previous prototypes.

An essential component of the interface is the Navigator, which relies on a
control panel. This panel allows users to combine spatial, temporal, and thematic selections, and specifies the information to be visualised at each moment. The
Navigator also gives access to the other components of the system (Visualiser,
Hyperbook, and Glossary). Users can select between maps and aerial

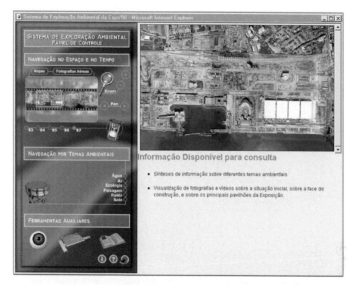

Fig. 7.6 Spatial Navigation Module: zooming and panning over aerial photos

Source: http://sig.cnig.pt/websea

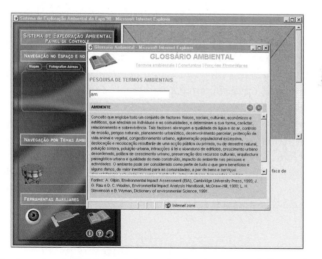

Fig. 7.7 Environmental Glossary: dynamic database search for environmental terms implemented with Dynamic HTML

Source: http://sig.cnig.pt/websea

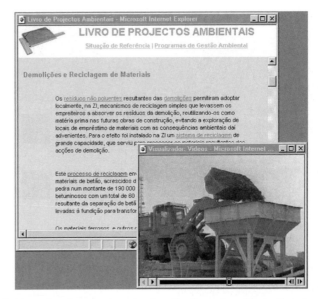

Fig. 7.8 Hyperbook component: visualisation of multimedia information associated with the Environmental Management Plan report

Source: http://sig.cnig.pt/websea

photographs as a basis for their spatial exploration. Typical navigation operations include zooming and panning. A timeline was also designed to enable users to visualise the temporal evolution of the project.

The Control Panel also allows thematic selections. The available themes correspond to the main environmental components and for each of them it is possible to visualise cartographic syntheses of the main problems and corresponding impacts. From these syntheses, one can access more detailed information available in the Hyperbook and Glossary.

Main problems of this version are associated with the different implementations of JavaScript and Dynamic HTML in Microsoft and Netscape browsers. This meant that code duplication was frequently necessary to implement the same functionality on these browsers.

7.3 Virtual Tejo

The development of decision support system tools for the Tejo Estuary began in 1985. Camara *et al.* (1990) reported the system Hypertejo based on the use of hypermedia to display relevant information and perform 'what . . . if' analysis for three types of users: decision-makers, technical staff, and the public. Virtual Tejo is a successor to Hypertejo: an experiment in the use of virtual reality as a front-end to a water quality management system.

Virtual Tejo included traditional databases, geographical information systems, and water quality simulation systems, but the interface was a virtual environment with a realistic representation of the terrain and the body of water of the Tejo Estuary stretch selected for the experiment. This stretch included the Expo98 intervention site.

Virtual Tejo had the following features (Camara *et al.*, 1998):

- There was a background of air, land, and water illustrated by three-dimensional models draped with photo-textures. Measures for environmental variables describing the background were available in a database.
- There was a set of artificial entities representing residential buildings, factories, and waste water treatment plants which were represented in a database.
- There was a set of natural entities including vegetation and fish represented by quasi-realistic models with an associated database.
- There was a set of water quality simulation models: non-point pollution, river quality, and estuarine quality. These models operated in a connected fashion.
- Users could manipulate entities (i.e., position waste water treatments, change land use, drive fish away). These actions activated the simulation models.
- Water pollution levels obtained by simulation were represented by textures for background draping or particles for water visualisation;
- Users could explore the estuarine basin by flying, walking, or diving into the water.

7.3.1 Architecture

Virtual Tejo relied on databases, a GIS server, and simulation models that were accessed through a virtual reality interface (Fig. 7.9). To integrate the different components, this project implemented a communication model based on an Object Linking and Embedding (OLE) approach. The GIS server accessed the databases through open database connectivity (ODBC).

The virtual reality interface launched the GIS and simulation servers. The simulation server also accessed the GIS server to use inputs for the models. The objects resulting from the simulations are accessed by the virtual reality client and used as input to visualisation (Camara *et al.*, 1998).

7.3.2 Environmental Simulation Models

The environmental simulation models included the estuarine model, which received inputs from models describing the river Trancao, and non-point pollution contributions. Data input for all models were set up on an interface based on

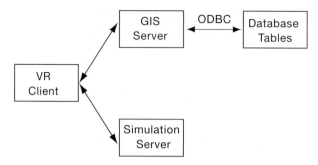

Fig. 7.9 Architecture of the virtual reality water quality management system
Source: http://virtual.dcea.fct.unl.pt/~jmr/sem97/vt/index.htm

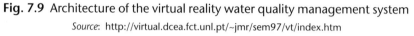

Fig. 7.10 Estuarine water quality model
Source: http://virtual.dcea.fct.unl.pt/~jmr/sem97/vt/index.htm

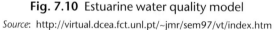

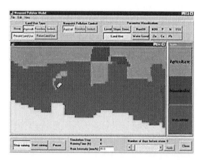

Fig. 7.11 River pollution model
Source: http://virtual.dcea.fct.unl.pt/~jmr/sem97/vt/index.htm

a sketching facility and boxes that were interactively filled with relevant data (Figs 7.10–7.12). It was also possible to insert/delete entities in the virtual reality interface and then input quantitative data in those boxes. Entities included streams, pollution sources, waste water treatment plants, and boundaries in the water body. Quantitative information was related to topographical, hydrological,

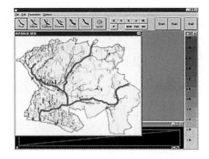

Fig. 7.12 Non-point pollution model

Source: http://virtual.dcea.fct.unl.pt/~jmr/sem97/vt/index.htm

and water quality data. The changes from three- to two-dimensional environments and back were easily achieved.

The estuarine model included two components: a hydrodynamic component and a water quality component. The hydrodynamic component was based on the solution using finite differences of the two-dimensional long-wave equations. It provided the values for the water level and velocity at each estuarine subdivision and simulation time instant.

The water quality component relied on a cellular automata formulation. Only biochemical oxygen demand (BOD) was considered, although the number of constituents could be extended. Cellular automata were used to track particles assuming three main processes: advection, dispersion, and decay. It provided a rapid visualisation of pollutant plumes.

The river quality model was based on the numerical solution of the traditional convection–diffusion model. BOD was the sole water quality parameter considered. Finally, the non-point pollution model considered a discretisation of the estuarine basin in a grid system and again a cellular automata formulation was used.

7.3.3 Virtual Reality Interface

User Interface Design

To design the user interface of Virtual Tejo, there was a priori identification of users, user goals, design goals, and user tasks. Users of a virtual reality water quality management system may include decision-makers, professionals from different disciplines, and the general public when involved in citizen participation exercises. Users may just want to browse the system or visualise the results of actions, either as external observers or participants in the virtual world. The design goals were to rely mainly on a realistic representation of the basin and visual operations and avoid an overload for the users. The user tasks

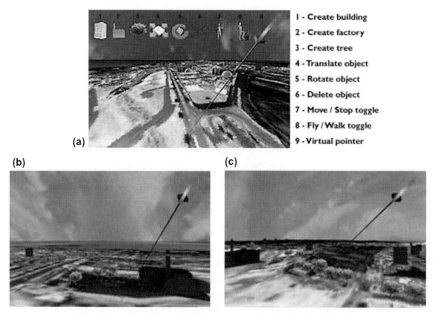

1 - Create building
2 - Create factory
3 - Create tree
4 - Translate object
5 - Rotate object
6 - Delete object
7 - Move / Stop toggle
8 - Fly / Walk toggle
9 - Virtual pointer

(a)

(b)

(c)

Fig. 7.13 Virtual Tejo user interface: (a) virtual pointer, objects, operators, and choice of navigation mode; (b) moving a waste water treatment plant; (c) changing a land use

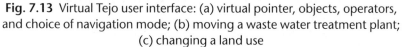

Source: http://virtual.dcea.fct.unl.pt/~jmr/sem97/vt/index.htm

were proposed to be of three types as outlined in Chapter 6: fundamental tasks related to problem perception; intermediate tasks involving navigation and visualisation; and advanced tasks that were related to data input to the simulation models.

The basic screen was the one represented in Fig. 7.13. It is based on a flight simulator screen with additional icons and without a flight panel. But as well as flying a basic navigation mode, it also allows walking (terrain following) and diving into the water. By dragging icons one adds or replaces entities in the virtual world. One can also eliminate entities. This is achieved by using the pointer. Note that the pointer acts as a sensor of environmental quality which is represented in a colour scale system, ranging from red (violating standards) to blue (acceptable measures).

Navigation

Navigation was implemented by means of the LOD algorithm proposed by Muchaxo *et al.* (1999). The user can navigate in either 'walk' or 'fly/dive' mode. In the 'walk' mode the viewpoint is maintained at a constant altitude relative to the terrain. In the 'fly/dive' mode there is no gravity, but there is still collision detection.

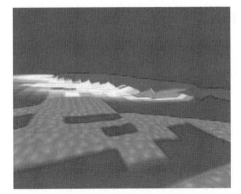

Fig. 7.14 Aerial view of water pollution in the virtual environment
Source: http://virtual.dcea.fct.unl.pt/~jmr/sem97/vt/index.htm

Fig. 7.15 Visualisation of pollutant plumes in the aquatic system
Source: http://virtual.dcea.fct.unl.pt/~jmr/sem97/vt/index.htm. Video available on the book's
Web site

Visualisation of Pollution

One can see the results of the simulation models using a raster representation
for the aerial view as in Fig. 7.14 or as particle plumes inside the water body as
in Fig. 7.15. The latter were visualised using particle rendering and fog effects
(see Chapter 5).

Pollution was also auralised. Surround sound was used to provide a global representation of the pollution level. Localised sound guided the users to the most
significant concentrations of pollutant particles. These were generated with the
cellular automata described above.

Fig. 7.16 Virtual wings

Parallel Experiments

During the Virtual Tejo, there were several parallel experiments, namely, the replication of animals and the development of browsing and note-taking facilities. The goal of developing replication was to offer the user the possibility of getting inside an animal in order to have its view of the world (generic fish and bird models were the two developed during the study). For this purpose, vision filters were developed as discussed in Chapter 5. For bird replication, artificial wings were developed. In their latest version they were centred in the use of sensors in the arms. One could then fly as in Fig. 7.16. The only purpose of the exercise was to experiment with the capabilities of VR systems. Later, some fashion designers used the concept in the development of flying jackets (see the book's Web site).

Browsing and taking notes in virtual environments was also already discussed in Chapter 6. The game Phantasmagoria offers similar capabilities. In this game, users can save the choices that have been made in the game and then view them in the end as a short movie (Korolenko, 1997).

There were also experiments that were planned but not executed due to limited resources. The most significant was a distributed interactive simulation project to test cooperative or independent water pollution control strategies of involved municipalities and industrial polluters.

7.4 Lessons Learned

There were many lessons learned from these two projects. They may be divided into two major groups: resources and management.

7.4.1 Resources

There are two key resources needed to develop interactive applications: computational and human. If interactive applications are to be used by the public, such as in the case of environmental digital documents, state-of-the-art computational resources are needed. For the older generation, the standard is television; for the younger, videogames. Both have a quality far superior to that of traditional technical software. These standards present a difficult challenge for virtual reality applications. Moore's law of computational power and, one hopes, a similar law for network bandwidth will alleviate this problem in the future. Interactive applications require teams that include (England and Finney, 1999):

- producer/director;
- programmers;
- designers;
- content experts;
- scriptwriters;
- video experts (director/producer, editor, journalist, and graphics artist);
- sound personnel (voice-over and sound editor);
- assistants.

Technically inclined producers tend to ignore a key aspect of such projects: they intend to tell a story. Scriptwriters are key members of an interactive movie team as they are for traditional movie-making. Unfortunately, the author's team did not have anyone with those abilities for both the Expo98 and Virtual Tejo projects, and clarity and attractiveness were lost.

There are two recent books on writing for multimedia (Garrand, 1997; and Korolenko, 1997), and include examples drawn from many experiences. Basically, there are three ways of telling a story:

- *The user as observer.* In this case, the user does not control characters but may select one plot path or another at predefined story branches. The user can then sample different viewpoints of the same story. An illustrative example is *Where in the World is Carmen San Diego?* (Korolenko, 1997).
- *The user as director.* Control of decisions, speech, or behaviour of one or more characters is made by the user such as in *King's Quest* (Korolenko, 1997).
- *The user as actor.* The user becomes a character as in *Myst* (Korolenko, 1997).

For environmental applications, the author believes that users may want to be observers, directors, and actors at different times:

- for the general browsing of a document, one wants to be an observer;
- a decision-maker wants to be a director;
- if a construction project affects one's neighbourhood, one may want to be inside the virtual representation and, in essence, become an actor.

This means that interactive environmental documents will have to be increasingly based on virtual reality representations coupled to simulation models as in Virtual Tejo, particularly for the user as a director or an actor. They will have to truly become interactive environmental movies and as movies they will need compelling scripts. Chapter 8 illustrates how interactive cinema may be used in future environmental impact assessment studies.

7.4.2 Management

Expo98's Environmental Exploratory System was largely a developmental project, although it had an underlying research component (Fonseca, 1998). Virtual Tejo was a research project. Both seem, thus, very different from a management standpoint. In a development project, time management is crucial and the final product may be reasonably defined from the start. But in interactive applications, there are always changes due to negative user reactions, copyright clearance difficulties, and technological breakthrough (that sometimes are radical leaps as in the explosion of the World Wide Web).

In a research project, such as Virtual Tejo, the goal is to experiment by breaking established limits and rules and living at the technological edge. Intermediate demonstrations and lack of funding usually set the deadlines.

Apart from these differences, there is always a constraint associated with multidisciplinary projects: the need for a common language. There is also, in managing any project, the goal of pushing all the team members to their limits and obtain a final product that is better than the sum of its components.

In more recent projects, the author has experimented with a tool that facilitates the achievement of these goals: the concept of the working site. The working site is the location where the project is presented, its time programming is established, and where all the ideas and advances are discussed by team members. It is also the place where all the news, publications, and other events related to the project are communicated. Finally, the working site is also a digital library of the multimedia data resources required for the project, relevant links, and other digital documents. The working site developed for the European Spatial Metadata Infrastructure (ESMI) project discussed in Chapter 3 (Box 3.1) illustrates the concept.

7.5 Further Exploration

The following sites are updated at (http://gasa.dcea.fct.unl.pt/camara/index.html). Student projects are also proposed at this site.

For a site introducing interactive movies, see Altman and Nakatsu's SIG-GRAPH tutorial (http://www.mic.atr.co.jp/events/siggraph97/course16/imovie/

imovie.html). Cartoon Network has Web Toons, which are illustrative interactive cartoons (http://cartoonnetwork.com).

For sites on interactive television, visit WebTV site (http://www.webtv.com). Open TV (http://www.opentv.com), and Interactive Television Report at (http://www.itvreport.com).

The Expo98 site on the environment may be accessed at (http://www.parquedasnacoes.pt/en/ambiente/default.asp).

The book's Web site includes videos on the three first Environment Expo98 prototypes. The final Web enabled CD-ROM may be found at (http://sig.cnig.pt/websea).

The Virtual Tejo slides and reports are available at (http://virtual.dcea.fct.unl.pt/~jmr/sem97/vt/index.htm). QuickTime videos are available on the book's Web site.

England and Finney (1999) cover the management of multimedia projects. The associated site is at (http://www.atsf.co.uk/manmult).

A general site on project management resources is Project Management Forum's at (http://www.pmforum.org).

References

Altman, E.J. and Nakatsu, R. (1997). Interactive Movies: Techniques, Technologies, and Content. *Course 16, SIGGRAPH 97*, ACM. New York: ACM Press available at (http://www.mic.atr.co.jp/events/siggraph97/course16/imovie/imovie.html).

Anderson, C. and Veljikov, M.D. (1990). *Creating Interactive Multimedia: A Practical Guide*. Glenville, IL.: Scott Foresman.

Camara, A.S., Silva, M.C., Carmona Rodrigues, A., Remedio, J.M., Castro, P., Fernandes, T., *et al.* (1990). Decision Support System for Estuarine Water Quality Management System. *Journal of Water Resources Planning and Management, ASCE*, 116/3: 417–32.

Camara, A.S., Neves, J.N., Muchaxo, J., Fernandes, J.P., Sousa, I., Nobre, E., *et al.* (1998). Virtual Environments and Water Quality Management. *Journal of Infrastructure Systems, ASCE*, 4/1: 28–36.

England, E. and Finney, A. (1999). *Managing Multimedia*. Harlow, UK: Addison-Wesley.

Fonseca, A. (1998). Spatial Multimedia Information Systems for Environmental Impact Assessment. PhD Dissertation, New University Of Lisbon, Monte de Caparica, Portugal.

Fonseca, A. and Camara, A.S. (1997). The Use of Multimedia Spatial Data Handling on Environmental Impact Assessment. In M. Craglia and H. Couclelis (eds), *Geographic Information Research: Bridging the Atlantic*, 556–69. London: Taylor & Francis.

Fonseca, A., Gouveia, C., Camara, A.S., and Silva, J.P. (1995a). Environmental Impact Assessment Using Multimedia Spatial Information Systems. *Environment and Planning B*, 22: 637–48.

Fonseca, A., Gouveia, C., Fernandes, J.P., Camara, A.S., Pinheiro, A., Aragão, D., *et al.* (1995*b*). Expo98 CD-ROM: A Multimedia System for Environmental Exploration. *Proceedings of the 1st Conference on Spatial Multimedia and Virtual Reality*, Lisbon, Portugal.

Fonseca, A., Gouveia, C., Fernandes, J.P., Camara, A.S., Pinheiro, A., Aragão, D., *et al.* (1999). The Expo'98 CD-ROM. In A.S. Camara and J. Raper (eds), *Spatial Multimedia and Virtual Reality*, 71–88. London: Taylor & Francis.

Garrand, T. (1997). *Writing for Multimedia*. Boston: Focal Press.

Ginige, A., Lowe, D.B., and Robertson, J. (1995). Hypermedia Authoring. *IEEE Multimedia*, 2/4: 24–35.

Korolenko, M.D. (1997). *Writing for Multimedia, A Guide and Sourcebook for the Digital Writer*. Belmont, CA: Wadsworth.

Muchaxo, J., Neves, J.N., and Camara, A.S. (1999). A Real-Time, Level-of-Detail Editable Representation for Photo-Textured Terrains with Cartographic Coherence. In A.S. Camara and J. Raper (eds), *Spatial Multimedia and Virtual Reality*. London: Taylor & Francis.

Shneiderman, B. (1989). Reflections on Authoring, Editing and Managing Hypertext. In E. Barret (ed.), *The Society of Text*, 115–31. Cambridge, MA: MIT Press.

8

Future Developments

8.1 Past and Future Visions

This book reviews methods for collecting, storing, modelling, visualising, and interacting with environmental data. Many of these methods are non-traditional, because of the multidimensional nature of the data. Environmental data are being increasingly derived from new sources of static and dynamic two- and three-dimensional imagery and other multimedia data such as sound.

It is also shown that the Internet is influencing the development of some of those methods and resulting tools. These, benefiting from the Internet protocols and standards, are becoming interoperable components that can be assembled for environmental problem solving.

Infrastructures of environmental information are also being created using the Internet. These infrastructures facilitate the work of environmental professionals and researchers. They also provide public access to environmental information enabling public participation in major projects.

The book focuses on spatial environmental problems that are typically approached in land use planning and environmental impact studies. The methods presented to address them may, however, be extended to non-spatial environmental problems.

Major environment concerns are now related to global problems. To address such problems, there are many new initiatives being proposed internationally. One of them, in line with the proposals contained herein, is Al Gore's vision of a Digital Earth. In Digital Earth (see the Web address in Section 8.8), one could fly-over the whole world represented by one-metre resolution satellite imagery and explore the associated database at different levels. The related TerraVision project team members (see Chapter 5 and the Web site address in Section 8.8) estimate that one million gigabytes are required to store the required information, which makes it unfeasible for the time being.

Progress in environmental knowledge can be advanced with new sources of multidimensional information, but fundamental problems remain to be solved. These problems hamper, most notably, the development of global climate, environmental–human health and ecosystem models.

One of the major hurdles stems from the millions of chemical products registered by the American Chemical Society (see the Web address in Section 8.8), with approximately 50,000 chemicals of those being used every day. The number of those created is in the order of thousands per week. Little is known about the effects of many new chemicals, as there is usually a delay between their introduction in the environment and the appearance of impacts on natural ecosystems and humans. Many chemicals remain in the environment for a long time and even if their short-term actions are known, ignorance remains on the long-term effects. There are also potential biomagnification and synergistic relationships that may not be readily anticipated. Continued chemistry and biology research is, however, necessary, even without ambitious projects such as the probable successor of the Human Genome project: the whole Biosphere Genome enterprise.

Research and development projects in computing, this book's main area of inspiration, may help in mitigating these and other environmental knowledge gaps. Environmental collaboratories, such as those described in Chapter 6, and infrastructures (introduced in Chapter 3), may contribute by organising existing data and pointing the direction to future research. The use of still-to-be-developed collaborative data models will facilitate the desirable seamless accessing of distributed databases associated to those digital laboratories and libraries (Greenspun, 1999).

There are other similar projects to pursue in the next few years (say until 2005), as discussed below. They will benefit from the continuing increase in computing power and improvements in wired and wire-less networks. By 2009, according to Kurzweil (1999), personal computers will process one thousand more instructions than the current desktops. Access to broadband networks (fixed or wire-less) will also be common well before that date using desktops or appliances that will enable ubiquitous computing (Cochrane, 1998).

Initiatives, such as Alliance (Smarr, 1997), will provide tools for having many of those projects implemented using *ad-hoc* supercomputers created by using the wasted cycles of Internet hosts (see Section 8.8 for the addresses of the Search for Extra-Terrestrial Intelligence and other related Web sites which are currently implementing these ideas). These supercomputers will allow realistic immersion in virtual ecosystems.

Many other computer science and telecommunications related developments that may help handling multidimensional environmental information are not reviewed here. Proposals for the next few decades have been discussed recently by leading innovators at ACM 97 (Denning and Metcalf, 1997) and at the thirty-fifth Anniversary of MIT's Laboratory for Computer Science (see the Web site addresses in Section 8.8). Other future-oriented computer surveys are offered by Dertouzos (1997), Cochrane (1998), Gershenfeld (1999*b*), and Kurzweil (1999).

The proposed developments are naturally extrapolations of past and current technologies and ideas. Many of these were proposed more than thirty years ago by Ivan Sutherland (see the Web site in Section 8.8) and other pioneers. Caution was preferred to risk, a path that even science fiction writers seem to have followed, as noted by Greenspun (1999). He mentioned, based on a communication by Eric Rabkin, Professor at University of Michigan that only in one case had a science fiction writer accurately predicted an invention: Arthur Clarke, who in 1945 suggested the possibility of geo-stationary satellites.

The suggested four areas for future research and development in environmental related sciences and professions are:

- Monitoring the natural and man-made environments by using augmented reality techniques, wearable computing, haptics, and environmental sensors.
- Creating environmental channels as front-ends for environmental agencies.
- Developing interactive movies for environmental impact assessment studies and urban and regional plans.
- Improving environmental education with a new generation of toys and games.

8.2 Augmenting Natural and Man-made Environments

Longley *et al.* (1999) refer to precision farming (Usery *et al.*, 1995) as an example of immersion in real environments. Precision farming enables, with the use of sensors (both remote and local), an innovative exploration of agricultural ecosystems. Its principles may be extended to the research of non-agricultural ecosystems. Additional tools may enhance the experience. These may include:

- Environmental sensors communicating in real time with laboratories.
- Augmented reality equipment including head-mounted displays, cameras, audio recorders, wearable computers, tablets, geographical positioning system receivers, wireless Internet access and (in the future) portable modelling tools based on haptics technologies.

The immersion may be complemented with earth observation imagery delivery. This imagery would provide a top view and may be used to guide the exploration. This environmental instrumentation will enhance traditional monitoring

Fig. 8.1 Augmenting the environment using wearable computers and portable modelling tools for estimating the volumes involved in land movements

Fig. 8.2 Augmenting the urban environment by superimposing maps of underground utility systems to real street images (inspired by an idea by Steve Feiner cited in Anders, 1999)

activities and may contribute to real ecosystem experiments as proposed by Carpenter *et al.* (1995). It will also contribute to develop the field of sensory ecology (see Chapter 5).

Key technologies to support this vision include environmental sensors (Chapter 2), augmented reality (Chapter 5), haptics tools (Chapter 5), wearable computing (Chapter 6), and wire-less access to the Internet (Chapter 6). Problems associated with these technologies were reviewed in those chapters.

Figures 8.1 and 8.2 show possibilities of future systems that will augment natural and built environments, respectively. In Fig. 8.1 the real image of a terrain before a land movement is compared to the generated image of the terrain after the movement. This movement is simulated by hand using a portable modelling tool (remember that this is a possibility). The system then computes immediately the volume of land involved. The same system may be used *in situ* by geologists to define formations in a three-dimensional visualisation of the terrain as proposed by Harbaugh (1993). Augmented reality systems may also facilitate classification of plants, soil, and animals by enabling the real-time accessing of multimedia databases.

Steve Feiner (cited by Anders, 1999) has planned the use of augmented reality systems to display maps of underground tunnels. The same idea may be applied to superimpose maps of underground utilities and other facilities on to the real image of a city street (Fig. 8.2).

Many of the utility maintenance problems occur in the older areas of cities. As GPS signals are either inaccurate or missing in such narrow street areas, correct positioning becomes impossible. Dias (1999) has investigated the use of radio or infrared emitters located in street network nodes to solve this problem. These emitters may send codes corresponding to predefined locations.

8.3 Environmental Channels

Environmental agencies around the world are developing infrastructures with data and software to support professionals. They are also mandated to provide public information on the state of the environment and promote the public's participation in environmental impact assessment studies. It will be a small step for some of the existing agencies to become true environmental channels on the Internet. The broadcast of environmental information is already made through The Weather Channel (see the Web address in Section 8.8) and most television and radio stations as discussed by Monmonier (1999). The Weather Channel includes forecasts on air pollution and allergy risks in major cities in North America.

Ubiquitous environmental sensors connected to the network may provide other real-time data on air pollution related to traffic, water pollution on a beach, and noise levels, for example. There will be increasing pressure from the public to have access to such information.

A strong Internet audience will also improve the chances of an environmental agency maintaining or increasing their governmental level of funding (a dangerous idea: imagine if all government agencies also wanted to become channels and compete for audiences on the Internet). To increase the number of visits to its site, the agency would have to upgrade their education and training programmes. It would also have to involve the public in monitoring, following the example of amateur astronomy (Finkbeiner, 1997) and the practice of the US EPA (see the Web site in Section 8.8).

The development of complaint systems would also facilitate public involvement. It is reasonable to anticipate that geo- and time-referenced complaint systems will become increasingly popular. Any individual could complain about noise, infrastructure maintenance, lighting, air, water, and solid waste problems. A verification system would ensure that most complaints will be legitimate. Complaint visualisation will become a major guiding source of information for non-routine monitoring activities. Other associated content could be Webcasts of augmented reality monitoring activities (Section 8.2), telepresence in scenic areas, and interactive environmental movies as discussed in Section 8.4.

Fig. 8.3 Entrance screen for the Portuguese Environmental Channel

Proposal developed by Infordesporto (http://www.infordesporto.pt) and Grupo de Analise de Sistemas Ambientais (http://gasa.dcea.fct.unl.pt). Available at http://gasa.dcea.fct.unl.pt/tec.old

Quality of content and the user interface will be essential for the creation of a successful environmental channel (see the proposal in Fig. 8.3 for Portugal's future Environmental Channel). However, its growth will depend on the development of communities of interested professionals, students, and the general public who want to learn and/or contribute. This development will depend on the efforts of the site maintainers and moderators. Greenspun (1999) shows that by using software packages some of their tasks could be facilitated (see the Web site in Section 8.8). Those packages help to keep:

- a database of users;
- a database of site content and how each piece relates to the others;
- a track of which users have browsed which pieces of content;
- a track of which external links users are selecting;
- a track of which users place and answer questions.

8.4 Interactive Environmental Movies

Environmental studies will produce reports, maps, images, video, and sound. However, some of the professional and research work and all the decision-making and citizen participation processes will be made using interactive environmental movies (see Chapter 7). Such movies will rely on the virtual ecosystem concept introduced in Chapter 5.

Virtual ecosystems will be housed in environmental information infrastructures. They will include realistic representations of ecosystems and associated databases and models. The fidelity will be similar to the current entertainment industry standards (see Dodsworth, 1998) by creating augmented reality representations mixing video with real-time generated realistic imagery (Barfield *et al.*, 1995).

Movies will run continuously, but may be interrupted at any time. Then they become interactive and enable three-dimensional exploration. This exploration will lead to actions that may trigger simulation models. The output of these models will be the subsequent stream of video. A quantitative decoding of all the actions and results will be available at any time during the interaction. For instance, by changing the land usage of a farm (with an area expressed in hectares), one will be able to estimate the average loading in suspended solids in tonnes per year resulting from non-point pollution.

Software to develop operational virtual ecosystems may be rented via the Internet. If necessary, in-house development will be facilitated by authoring tools that will assemble available components. Acquisition of information via imagery and programming by reproduction will also contribute to rapid prototyping of simulation models. These models will provide realistic three-dimensional interactive visions of the ecosystem evolution (based on developments discussed in Chapter 4).

The public and decision-makers will play with interactive movies as they do with games. Interactive movies representing impact statements and plans will be designed to have plays (i.e., the control variables that describe the project or the plan) and will generate results for those plays (via the key impact variables associated to the project or plan). They will include interactive three-dimensional visualisation capabilities, two-, and three-dimensional interaction tools.

The study reports will be a complementary source of information and interaction. A common causal diagram will connect scripts and associated interactive movies. This diagram will enable changes in the script to be visualised in the interactive movie. Professionals and researchers may also change directly model parameters and initial values. The interactive movie representation will immediately adjust to those alterations (for an example of linked software in ecological design environment see DOME's Web site listed in Section 8.8).

Interactive movies will be available on the Internet and may be used collaboratively by groups of individuals. To facilitate this collaboration process, people will have the successor of the Clear Board system available (see Chapter 6). The Internet consultation will provide the basis for defining interest groups (those who typically select the same plays belong to the same group). Special settings

Fig. 8.4 Installation for a public meeting using a mural that can include multiple screens. Alternative projects are run in the central four screens as interactive movies. Voting is processed in the lateral screen at the left. The comparative analysis of the impacts of the different alternatives may be visualised in the lateral screen on the right

could be devised to facilitate the communication of the views of the different interest groups in public meetings. Those may include shared screens and work benches as discussed in Chapter 6.

Public meetings may also rely on innovative installations such as that inspired by Karl Sims's Galapagos (see the Web site address in Section 8.8):

- The room would have multiple screens, each representing a replay of the interactive movie experience of each interest group. The replays could be divided into stages (i.e., in the impact assessment of an infrastructure, there are typically before, during, immediately after and stable phases in the implementation process).
- Public voting could then be done in succession to eliminate the most unreasonable interest group proposals, cluster close views, and ultimately shape the most satisfying project alternative from an environmental standpoint. Both the voting procedures and any proposal resulting from a combination of views from different interest groups could be readily displayed in a mural.

Figure 8.4 shows the proposed installation. Section 8.8 includes Web addresses of sites describing successful art installations that may be used as an inspiration to design public meeting places at the beginning of the third millennium.

8.5 Environmental Education

Traditional field based environmental education activities will continue. Chemistry sets, microscopes, and telescopes will continue to be popular.

Fig. 8.5 Supersonic Ear, an example of an augmented reality toy

Source: http://www.wildplanet.com. Courtesy of Wild Planet Toys

However, environmental toys and games will become key components of environmental education and will use many of the concepts described in this book.

Toys are usually reduced models of real objects (i.e., toy car). Future environmental toys will include sets of components that will enable the development of simple sensors. They will also include basic augmented reality outfits to explore ecosystems. In fact, Lego Mindstorms and some of the Wild Planet toys (see Section 8.8 for their Web site addresses) seem to point in that direction. Figure 8.5 shows an environmental toy from Wild Planet.

SimCity 3000 is an excellent example of an educational game (see the Web site address on SimCity 3000 in Section 8.8). SimCity represents today what professional plans will be in the near future. With additional processing power and bandwidth, it can only improve. Another remarkable example of what an environmental education game should be is the Virtual Fish Tank referred to in Chapter 4.

There are many possible paths for environmental game development using virtual reality capabilities which are still unexplored because of computer and other hardware and telecommunications limitations. The capabilities of traditional virtual environments (i.e., fly-over, object manipulation, scaling), distributed interactive simulations (i.e., interaction among users in distributed virtual worlds), and haptics (i.e., force feedback and touch) implemented together do open the possibility for compelling new environmental education games.

8.6 Preparing Future Environmental Professionals

The developments reported in this book are dramatically changing the methods used to collect, process, and report environmental information. Topics, such as database backed Web site development, image processing, three-dimensional modelling, animation, visualisation, and even sound processing, will increasingly become required for the training of environmental professionals.

Those topics may be covered at different levels ranging from actually teaching students how to develop some of the tools to merely training them on their use. The author believes in teaching (and learning) traditional programming and is biased towards the first option. The open-ended projects, which are recommended on the book's Web site, reflect that bias. The preference results from answering a simple question: given a student that knows how to develop and another that only knows how to use, which student should an employer hire?

There are many degree courses in engineering and science with an emphasis on teaching how to develop. Web site addresses for illustrative examples are provided in Section 8.8.

Other topics that deserve consideration in designing curricula for the next generation of environmental professionals are traditional: language and typography. Text is still the dominant data type on the World Wide Web. It is by the written language that arguments are presented. Type may be used to convey emotions (see Chapter 5) or improve the aesthetics of a site. Unfortunately, language and typography are neglected in engineering and science degrees.

Finally, future environmental professionals should be prepared to respond to the needs of different types of employers, but also to the challenges of creating their own companies. Opportunities provided by some of the technologies presented in this book are numerous. For an entertaining and useful guide to create new businesses see Kawasaki (1999) and his Garage Web site (Section 8.8).

8.7 A Final Cautionary Comment

Even a list of conservative proposals for future developments may be off the mark. Corn and Horrigan (1984), with their remarkable review on past visions of the American future, showed how even perfectly plausible designs did not succeed.

8.8 Further Exploration

The list of the following sites is periodically updated at (http://gasa.dcea. fct.unl.pt/camara/index.html).

Future Visions

To read Al Gore's Digital Earth speech see (http://www.regis.berkeley.edu/ rhome/whatsnew/gore_digearth.html). The TerraVision system Web site is (http://www.ai.sri.com/TerraVision/TV-II/index.html).

Browse the American Chemical Society site (http://www.acs.org) to find out the daily increase in the number of chemical products.

Idle computing resources on the Internet are being used in the Search for Extra-Terrestrial Intelligence (SETI) projects (http://setiathome.ssl.berkeley .edu/). About 400,000 people participate by donating their computers to run a program that downloads and analyses radio telescope data. The attraction is 'the possibility of detecting a faint murmur of a civilisation beyond Earth', according to the SETI Web site. A site reviewing similar computing initiatives is the Peer-to-Peer Working Group Web site (http://www.peer-to-peerwg.org).

To read predictions on the future computing methods see:

- Ivan Sutherland's views (http://www.sun.com/960710/feature3/).
- ACM 97 The Next 50 Years, a collection of views from leading experts in computer science (http://research.microsoft.com/acm97).
- MIT's Laboratory for Computer Science 35th Anniversary Webcast (http://www.lcs.mit.edu).
- Projects on High Performance Computing at http://alliance.ncsa.uiuc.edu.

Augmenting Natural and Man-made Environments

The leading site in augmented reality is at (http://www.cs.rit.edu/~jrv/research/ar). Information on wearable computing may be found at (http://wearcam.org). Sensable (http://www.sensable.com) technology may be used to model landscapes and terrains manually in workstation environments.

An excellent review site on smell, touch, taste, sensors (specially biosensors and chemical sensors), GPS, telepresence and haptics is Stephen Wilson's Conceptual Information Arts (http://userwww.sfsu.edu/~swilson/emerging/wilson.newtech.html).

Unconventional displays are provided by companies such as Ultimate Display (http://www.ultimatedisplay.com).

Third generation mobile phones will provide a new platform for augmented reality as discussed at (http://www.w3.org/Mobile/posdep/oulu.html). Sites on third generation mobile phones may be found at (http://www.nokia.com/3g/index.html), (http://www.ericsson.com/mobileinternet), and (http://www.umts-forum.org).

Environmental Channels

The Weather Channel is an illustrative example of an environmental channel (http://www.weather.com). To browse The Environmental Channel prototype, see (http://gasa.dcea.fct.unl.pt/tec.old). Existing environmental channels include Environmental News Network (http://www.enn.com) and Vertical Net sites such as (http://solidwaste.com). The latter offers auction activities.

Sites related to public participation in environmental monitoring include (http://www.globe.gov) and (http://www.epa.gov/epahome/citizen.htm).

Philip Greenspun's ArsDigita site (http://www.arsdigita.com) shows how to build and maintain virtual communities.

Interactive Environmental Movies

MIT's DOME software for ecological design shows how different applications can operate in concert (http://cadlab.mit.edu/).

Karl Sims' Galapagos is introduced at (http://www.ntticc.or.jp/permanent/karl/karl_e.html).

A list of links to inspiring sites covering interactive cinema basic technologies is located at Stephen Wilson's (http://userwww.sfsu.edu/~swilson/emerging/wilson.newtech.html).

Christa Sommerer and Laurent Mignonneau have created unusual interactive experiences. See their Web site (http://www.mic.atr.co.jp/~christa).

Environmental Education

To learn from popular toys and games see:

- Wild Planet toys (http://www.wildplanet.com).
- Lego Mindstorms (http://www.lego.com).
- SimCity 3000 (http://simcity.ea.com).

Preparing Future Environmental Professionals

There are numerous courses stressing how to understand and build tools rather than just use them. Examples include Neil Gershenfeld's How to Make Almost Anything and The Nature of Mathematical Modelling (http://www.media.mit.edu/physics/pedagogy) and Philip Greenspun's Software Engineering of Innovative Web Services (http://6916.lcs.mit.edu).

To start a company, see http://www.garage.com. Red Herring provides a review on emerging companies in new technologies (http://www.redherring.com).

References

Anders, P. (1999). *Envisioning Cyberspace*. New York: McGraw-Hill.

Barfield, W., Rosenberg, C., and Lotens, W. (1995). Augmented-Reality Displays. In W. Barfield and T.A. Furness III (eds), *Virtual Environments and Advanced Interface Design*, 542–76. New York: Oxford University Press.

Corn, J.J. and Horrigan, B. (1984). *Yesterday's Tomorrows, Past Visions of the America Future*. Baltimore, MD: The Johns Hopkins University Press.

Carpenter, S.R., Chisholm, S.W., Krebs, C.J., Schindler, D.W., and Wright, R.F. (1995). Ecosystem Experiments. *Science*, 269: 324–46.

Cochrane, P. (1998). *Tips for Time Travellers*. New York: McGraw-Hill.

Denning, P.J. and Metcalfe, R.M. (1997). *Beyond Calculation: The Next Fifty Years of Computing*. New York: Copernicus.

Dertouzos, M. (1997). *What It Will Be*. New York: Harper.

Dias, A.E. (1999). Contextual Awareness in the Exploration of Spatial Systems. PhD Dissertation, New University of Lisbon, Monte de Caparica, Portugal.

Dodsworth, C. (ed.) (1998). *Digital Illusion: Entertaining the Future with High Technology*. Reading, MA: Addison Wesley.

Finkbeiner, A. (1997). The Dial-Up Sky. *Science*, 278: 1010–12.

Gershenfeld, N. (1999a). *The Nature of Mathematical Modelling*. Cambridge, UK: Cambridge University Press.

Gershenfeld, N. (1999b). *When Things Start to Think*. New York: Henry Holt.

Greenspun, P. (1999). *Philip and Alex's Guide to Web Publishing*. San Francisco: Morgan Kauffman (available at http://www.arsdigita.com).

Harbaugh, J. (1993). Virtual Reality and Geology. Personal communication, Panel on Future Developments, *ESF Conference on 3D Modelling of the Natural Unbounded Domain*, Il Ciocco, Italy.

Kawasaki, G. (1999). *Rules for Revolutionaries: The Capitalist Manifesto for Creating and Marketing New Products and Services*. New York: Harper.

Kurzweil, R. (1999). *The Age of Spiritual Machines*. New York: Viking (book site at http://www.penguinputnam.com/kurzweil).

Longley, P.A., Goodchild, M.F., Maguire, D.J., and Rhind, D.W. (1999). Epilogue: Seeking Out the Future. In P.A. Longley, M.F. Goodchild, D.J. Maguire, and D.W. Rhind (eds), *Geographical Information Systems*, Vol. 2. New York: John Wiley.

Monmonier, M. (1999). *Air Apparent*. Chicago: University of Chicago Press.

Smarr, L. (1997). Toward the 21st Century. *Communications of the ACM*, 40/11: 29–32.

Usery, E.L., Pocknee S., and Boydell, B. (1995). Precision Farming Data Management Using Geographic Information Systems. *Photogrammetric Engineering and Remote Sensing*, 61: 1383–91.

Appendix 1

World Wide Web Tools and Technologies

This appendix provides an overview of sites that may help the reader in developing, marketing, and browsing Web sites. A survey of key protocols and standards is also included.

A1.1 Developing Web Sites

General Sites

There are several sites that provide comprehensive access to tools. Three examples are the Webmonkey, Webdeveloper, and the Web Developer's Virtual Library at, respectively (http://www.hotwired.com/webmonkey), (http://www.webdeveloper.com), and (http://www.wdvl.com).

For electronic versions of books on the Web see the O'Reilly site (http://www.oreilly.com).

HTML

Developments in the Hypertext Markup Language (HTML) are covered by the World Wide Web Consortium (W3C) (http://www.w3.org). Muscione and Kennedy (1998) provide a comprehensive guide on HTML.

XML

Goldfarb and Prescod (1998) introduce The Extensible Markup Language (XML). XML is discussed on the W3C and also at (http://www.xml.com).

VRML

Ames *et al.* (1997) is a reference book on VRML 2.0. There are many working groups extending the capabilities of VRML 1997. These developments are reported at (http://www.web3d.org).

Graphics on the Web

Lynda Weinman's (1999) book site provides helpful information. See (http://www.lynda.com) for information and relevant links.

Murray and vanRyper's (1996) book site has information on graphics file formats for the Web (http://www.ora.com/centers/gff/) such as GIF, JPEG, and Portable Network Graphics (PNG).

Video, Audio, and Animation

For an overview on video, audio and animation in the Web see Naik (1998), Wagstaff (1999), and Beggs and Thede (2001). Sites on different formats and tools are included below:

- QuickTime and QuickTime VR (http://www.quicktime.apple.com).
- Windows Media movies (http://www.microsoft.com).
- MPEG (http://www.cselt.it/mpeg).
- Sites on audio include (http://www.real.com) and (http://www.vocaltec.com).
- A Web site on GIFs and animated GIFS is available at (http://member.aol.com/royalef/gifanim.htm).
- Shockwave/Flash site (http://www.macromedia.com).

Java

The Sun's Java page (http://java.sun.com) provides a comprehensive coverage on the language's related developments.

JavaScript

For JavaScript related documents see (http://developer.netscape.com/one/javascript).

Active X

Active X usually means Active X controls and Active X scripting. Active X controls may be coded in a variety of languages including Java (see above) and run within browsers, word processors, or spreadsheets (see inovaGIS in Chapter 3). Active X scripting (that allows code to exist in HTML files) consists of two technologies: an Active X scripting host, which can be a browser, a server, or other application; and an Active X scripting engine that interprets the script, which may be in Perl, VBScript, or JavaScript (see above). Information on Active X is available at the Microsoft site (http://www.microsoft.com).

Database Backed Web Sites

Greenspun (1999) provides an overview on Web sites backed by databases at (http://www.arsdigita.com).

For the use of Active Server Pages, the method referred to in Chapter 3, see (http://www.microsoft.com).

For recent developments on the use of Common Gateway Interface (CGI), an inefficient method to access databases (as it requires a new process to start each time a page is served), see (http://www.fastcgi.com). The lasting popularity of CGIs despite their short-comings relies on the simplicity of its development and portability.

For information on the PHP scripting language see (http://www.php.net).

Wire-less Internet

Wire-less Internet services are provided in the United States using radio technology by (www.ricochet.net).

The Internet is available for mobile phones using the Wireless Application Protocol (WAP), see (http://www.wapforum.org) and (http://gelon.net).

World Wide Web Guidelines

Sites that provide guidelines on the design of Web sites include:

- IBM's design guidelines
  (http://www.ibm.com/ibm/hci/guidelines/web/web_design.html).
- Human–Computer Interaction Virtual Library (http://usableweb.com).

Sammons (1999) offers a practical guide on how to write for Web sites.

Authoring Tools

To facilitate the edition of Web sites you can use a number of authoring tools, which may start by simply using a word processor and saving the document in HTML. Fully fledged popular authoring tools include:

- Netscape Composer (http://home.netscape.com).
- Microsoft Front Page (http://www.microsoft.com).
- Adobe's GoLive (http://www.adobe.com).
- Macromedia's Dreamweaver (http://www.macromedia.com).

For reference to many other existing authoring tools see (http://www.hotwired.com/webmonkey).

A1.2 Marketing Web Sites

Greenspun's book includes tips on marketing a Web site (Greenspun, 1999). Dyson's work provides insights on the role of Internet services (Dyson, 1997).

Before marketing a site, one has to register the domain at (http://rs.internic.net). Net4India (http://www.net4india.com) is an alternative site for domain registration.

For a general site on marketing Web sites see (http://www.wilsonweb.com/webmarket).

One can use the following services, on a limited basis, to submit a site to a search engine:

- Submit–it (http://siteowner.linkexchange.com/Free.cfm), only 20 locations.
- NetAnnounce (http://netannounce.com/freelist.html?s=netannounce.com/ gold.html).
- Siteregister (http://www.siteregister.com/sample.submissions-main.html).

Paying a fee you can have the site submitted to hundreds of sites in:

- Submit–it (http://submitit.linkexchange.com).
- WebPromote (http://www.webpromote.com).

The site may also be announced in user groups. There are more than 20,000 user groups. Two relevant for the topics covered in this book are (News:comp.infosystems.gis–GIS) and (News:sci.image.processing).

A free of charge method to measure the number of visitors to the site is the Nedstat service, available at (http://be.nedstat.net/fr/info.html). Customised Web metrics services are provided by firms such as (http://www.mediametrix.com) and (http://www.i33.com).

A1.3 Browsing Web Sites

To find a Web site you may use a search engine. To view a multimedia site you may have to download plug-ins.

Search Engines

Popular search engines include:

- Altavista (http://www.altavista.com).
- Lycos (http://www.lycos.com).
- Magellan (http://www.mckinley.com).
- Excite (http://www.excite.com).
- Yahoo (http://www.yahoo.com).
- Webcrawler (http://www.webcrawler.com).
- Infoseek (http://www.infoseek.com).
- Hotbot (http://www.hotbot.com).
- Google (http://www.google.com).

Simultaneous use of search engines is provided by:

- Metacrawler (http://www.metacrawler.com).
- Metasearch (http://www.metasearch.com).
- AskJeeves (http://www.askjeeves.com).

Key Plug-ins

Key plug-ins that one should download include:

- Adobe Acrobat Reader (.pdf) (http://www.adobe.com).
- QuickTime player (.mov) (http://www.quicktime.com).
- Real player (.rm, .ra) (http://www.real.com). Real player relies on streaming technology to view and listen to video and audio files as they download to one's computer.
- Windows Media (http://windowsmedia.com).
- Winzip (.zip)/Stuffit Expander (.sit, .hqx) (http://www.winzip.com), (http://www.aladdinsys.com/expander). Applications for Windows and Mac OS that can compress and decompress files.
- Winamp/Macamp (.MP3) (http://www.winamp.com), (http://www.macamp.com). These plug-ins are audio players for Windows and Macintosh.
- Shockwave and Flash (http://www.macromedia.com). These plug-ins are viewers for interactive Web sites and animations.
- Cosmo Player (http://www.sgi.com). Cosmo is a viewer for VRML files.
- WHIP! (http://www.autodesk.com/products/whip/index.htm). WHIP is a viewer for DWF (Drawing Web Format).
- Volo View Express (http://www.autodesk.com/products/volo/index.htm). Volo is a viewer for DWG and DXF files, two popular formats in Computer Aided Design.

A1.4 Key Internet Protocols and Standards

The field of networking is divided into a seven-layer data communications model by the International Organization for Standardization Open Systems Interconnection (ISO/OSI):

- *Physical layer*: describes physical properties of any media and the signals that transport information.
- *Data-link layer*: transports data across the physical media. This layer deals with physical addresses such as those used by Ethernet.
- *Network layer*: related to routing between nodes in a network. This layer uses logical addresses. Relevant protocols include the Internet Protocol (IP) for addressing nodes and the Internet Control Message Protocol (ICMP).
- *Transport layer*: provides services similar to the network layer. Its major function is to ensure reliability. The Transmission Control Protocol (TCP) and the User Datagram Protocol (UDP) are protocols associated with this layer.
- *Session layer*: enables the establishment, management, and disconnection of sessions. Sessions exist when a connection between two points is made (an example provided by Naik, 1998, is the telephone call). Simple Mail Transport Protocol (SMTP), File Transfer Protocol (FTP), and Telnet are protocols relevant to this layer.
- *Presentation layer*: provides the means to facilitate operations such as mail exchanges or file transfers.
- *Application layer*: uses the services of the *presentation layer* to execute the applications referred to above.

The Internet relies on several devices associated with these layers such as (in a simplified view):

- *Repeaters*, operating at the physical layer to extend the signal.
- *Bridges*, working at the data-link layer to connect Local Area Networks.
- *Routers*, operating at the network layer that connect islands of networks.
- *Hosts*, including computers connected to form networks, with some of these forming islands.

The World Wide Web relies on the Hypertext Transfer Protocol (HTTP). HTTP specifies how a Web browser asks for a document from a Web server.

The Multi-Purpose Internet Mail Extensions (MIME) is the protocol that enables the specification of the types of documents to be served by an HTTP server such as text, image, audio, and application. It was initially developed to enable the use of images and other non-text documents in e-mail messages (Greenspun, 1999).

A Uniform Resource Locator (URL) locates Internet documents. URLs may follow the HTTP protocol, FTP and News protocols, and encrypted protocols such as HTTPS.

A1.5 *References*

Ames, A.L., Nadeau, D.R., and Moreland, J.L. (1997). *VRML 2.0 Sourcebook*. New York: John Wiley.

Beggs, J., and Thede, D. (2001). *Designing Web Audio*. Sebastopol, CA: O'Reilly.

Dyson, E. (1997). *Release 2.0*. New York: Broadway Books.

Goldfarb, C.F., and Prescod, P. (1998). *The XML Handbook*. Upper Saddle River, NJ: Prentice Hall.

Greenspun, P. (1999). *Philip and Alex's Guide to Web Publishing*. San Francisco, CA: Morgan Kauffman (see also http://www.arsdigita.com).

Murray, J.D., and vanRyper, W. (1996). *Encyclopedia of Graphics File Formats*. Sebastopol, CA: O'Reilly (see also http://www.ora.com/centers/gff/).

Muscione, C., and Kennedy, B. (1998). HTML, *The Definitive Guide*. Sebastopol, CA: O'Reilly (see also http://www.oreilly.com).

Naik, D.C. (1998). *Internet Standards and Protocols*. Redmond, WA: Microsoft Press.

Sammons, M.C. (1999). *The Internet Writer's Handbook*. Boston, MA: Allyn and Bacon.

Wagstaff, S. (1999). *Animation on the Web*. Berkeley, CA: Peachpit Press (see also http://www.peachpit.com/ontheweb/animation).

Weinman, L. (1999). *Designing Web Graphics: 3*. Indianapolis, IN: New Riders.

Appendix 2

Image Processing Primer

A2.1 Introduction

Image processing refers to the manipulation of a digital image in order to extract more information than the one currently visible. A digital image is composed of a matrix of picture elements, called pixels. The resolution of an image is considered to be the dimensions (number of lines × number of columns) of the image matrix.

Digital images may also be seen as displays of a discrete set of brightness points. The brightness of a pixel is the intensity of the image at that point and results from two components: the source of light on the scene being viewed (illumination); and the amount of light reflected by the objects in the scene (reflectance).

The brightness value at each pixel location is expressed in terms of the number of bit planes. This value is a number ranging from 0 (black) to 255 (bright white), when the number of bit planes is 8 (each bit corresponds to two colours: black and white; the number of combinations is thus 2 to the power of 8, or 256 grey levels). A colour image might have 8 bit planes for each of the three red, green, and blue colour channels (RGB).

When digitising an image, there are two major issues: sampling and quantisation (Gonzalez and Woods, 1992). Sampling means choosing those points needed to represent a given image. Sampling is mapping an image from a continuum of points in space into a discrete set, when considering an analogue image. Given a digital image, sampling is mapping from one discrete set of points to another smaller set. Quantisation is related to the amplitude of grey levels required.

In image processing, manipulation can be carried out not only on the original array of pixels but also on the image's spectral representation. The first type of operation is said to be in the spatial domain. The latter refers to operations carried out with Fourier transforms. These are obtained by separating information about an image according to the spatial frequencies of features such as discontinuities, gradients, or background shading. Operations with the matrix of numbers of the Fourier-transformed image are said to be in the frequency domain.

For decomposing an image into different resolution scales, the use of multiscale transforms, such as wavelet functions, are appropriate. They enable obtaining multilayered views of reality and facilitate the real-time editing of images, as discussed by Muchaxo and Neves (1999).

A2.2 Spatial Domain

Image transformations are called filters. The simplest of these are operations at the pixel level (or point operations), which are essentially grey level (or colour) transformations. These operations are used for contrast enhancement, where for instance, pixels with values below a certain level would be brightened or darkened.

A histogram (see Fig. 2.6), is a convenient method of representing the distribution of pixels across grey levels. It has also been shown that the manipulation of histograms, such as stretching the pixel distribution across grey levels or distributing it more evenly from black to white (histogram equalisation), can help to bring more detail to the image. The advantage of equalisation over stretching is its automatic nature (in the former, one has to hand pick the extreme points) (Wolff and Yaeger, 1993).

Image processing filters may also be based on window operations performed on small $m \times n$ matrices (i.e., 3×3, 5×5, 7×7), which are called 'masks' or 'kernels'. Window operations involve computation of individual pixel values with respect to the brightness values of the pixels in the neighbourhood. They are performed by moving the mask from pixel to pixel throughout the image, in a process which is also called 'convolution filtering'.

Given a window of pixels (Wolff and Yeager, 1993):

$$I_{ij} = \begin{bmatrix} P_{11} & P_{12} & P_{13} \\ P_{21} & P_{22} & P_{23} \\ P_{31} & P_{32} & P_{33} \end{bmatrix}$$

And a mask:

$$M_{ij} = \begin{bmatrix} M_{11} & M_{12} & M_{13} \\ M_{21} & M_{22} & M_{23} \\ M_{31} & M_{32} & M_{33} \end{bmatrix}$$

Then for the pixel P_{22}, the new value becomes:

$$P_{22} = M_{11}P_{11} + M_{12}P_{12} + M_{13}P_{13} + M_{21}P_{21} + M_{22}P_{22}$$
$$+ M_{23}P_{23} + M_{31}P_{31} + M_{32}P_{32} + M_{33}P_{33}$$

Examples of convolution filtering include:

- *Low-pass filtering.* The brightness value of a given pixel and the values of the neighbouring pixels are evaluated, and a mean of this convolution is obtained (coefficients of matrix M_{ij} are all equal to 1). The process is then repeated for each pixel. The result is that high spatial frequency detail is smoothed out (and thus the low-pass label). As blurring may occur, weighted means have been computed to preserve the edges (coefficients of M_{ij} may differ from one another).
- *Median filtering.* In this case, each pixel value is replaced by the median of its neighbours (the value such that 50% of the values in the neighbourhood are above and the remaining 50% are below). This method is generally good at preserving edges.

- *Mode filtering*, where each pixel value is replaced by its most common neighbour. This filtering method is suitable for classification procedures where each pixel corresponds to an object, which must be placed into a class.
- *High-pass filtering*, which is applied to enhance the high-frequency local variations. Sharpening the edges may be one of the goals. This is achieved by applying convolution masks where all the M_{ij} coefficients are equal to -1 and M_{22} is equal to 9 (Jensen, 1996).

A2.3 Frequency Domain

Frequency domain analysis is based on the separation of an image into its spatial frequency components. This is achieved by applying the Fourier theorem that states that any function $f(x)$ can be represented by a summation of a series of sinusoidal terms of varying spatial frequencies k (also called wavenumbers), where:

$$k = 2\Pi/\lambda$$

and λ is a wavelength.

The Fourier transform of a function $f(x)$ is

$$F(k) = \int_{-\infty}^{\infty} f(x)e^{ikx}dx$$

where x is a spatial variable and i is equal to $\sqrt{-1}$.

The function $f(x)$ can be determined from $F(k)$ by the inverse Fourier transform:

$$f(x) = \int_{-\infty}^{\infty} F(k)e^{-ikx}dk$$

To apply Fourier analysis to image processing these equations need to be expanded into two dimensions and approach the images as being matrices. The appropriate equations are then:

$$F(k,1) = 1/MN \sum_{k=0}^{M-1}\sum_{1=0}^{N-1} f(x,y)e^{i(kx/M+1y/N)}$$

for the two-dimensional discrete Fourier transform. And

$$F(x,y) = \sum_{k=0}^{M-1}\sum_{1=0}^{N-1} F(k,1)e^{-i(kx/M+1y/N)}$$

for the inverse Fourier transform, where:

- N is the number of pixels in the x direction and M, the number of pixels in the y direction; and
- 1 is the wavenumber of variable y.

$F(k,1)$ contains the spatial frequency information of the original image $f(x,y)$. To display it $F(k,1)$ is used as a sum of the real part $R(k,1)$, and an imaginary part $iI(k,1)$, and then apply:

$$|F(k,1)| = \sqrt{R(k,1)^2 + I(k,1)^2}$$

where $|F(k,1)|$ represents the magnitude and direction of the different frequency components in the image $f(x,y)$.

The filters applied in the spatial domain can also be used in the frequency domain. This is done by multiplying the Fourier transform of the original image by a mask called a 'frequency domain filter'. Then, by computing the inverse Fourier transform of the manipulated frequency spectrum, a filtered image is obtained in the spatial domain (Jensen, 1996). Frequency domain filters can be used to remove frequency components of the image which may not be easily removed with spatial domain filters.

An important Fourier transform relationship is convolution. The convolution of two functions $f(x)$ and $g(x)$ is:

$$f(x) * g(x) = \int_{-\infty}^{\infty} f(\tau) g(x - \tau) d\tau$$

where τ is an integration parameter. Wolff and Yaeger (1993) state that if $f(x)$ is the signal and $g(x)$ is a filter in space, the equation indicates that the value of the convolution at each point, x, is the integral of the product $f(x)$ with the filter $g(x)$. This filter is moved so that its origin is at x. Extending to two dimensions and applying to image filtering, the convolution operation is like having the spatial filter moving around the image.

The convolution of f and g can be determined by computing the product of the Fourier transforms of f and g, that is:

$$f(x,y) * g(x,y) = \sum_{k=0}^{M-1} \sum_{l=0}^{N-1} f(k,l) g(x-k, y-1)$$

where M and N are the number of pixels in the x and y directions (Gonzalez and Woods, 1992; Wolff and Yaeger, 1993).

Convolutions facilitate filtering in the frequency domain, as one just multiplies the filter with the Fourier transform of the image. However, the computation of the summations of the equation above is of the order N^2, where N is the number of points (Gonzalez and Woods, 1992). The use of fast Fourier transforms (FFT) instead of ordinary Fourier transforms reduces the computation time by several orders of magnitude, as FFTs reduce from N^2 to $N \log N$ the size of the calculations involved. Gonzalez and Woods (1992) and Press *et al.* (1992) provide the algorithms to compute an FFT.

A2.4 Wavelets

The Fourier transform maps the original input data onto a new space, where the basic functions (sine functions) extend from plus- to minus-infinity. The wavelet transform also maps input data on to a new space. However, the basis functions of this space are defined as being a limited, finite domain, features that cause wavelets to be termed 'compact supports'. This compactness demonstrates that wavelet transforms are more efficient for computing than even the fast Fourier transforms: wavelets have a complexity in the order of N, while FFTs have a complexity in the order of $N\log N$ (Starck *et al.*, 1998).

There are many wavelet sets that can be selected depending on the problem. Problems addressed advantageously by wavelets (over the use of Fourier transforms) are those having local scale-related features. The site on wavelets listed below provides access to sites that describe several discrete wavelet transforms (DWTs) and associated inverse wavelet transforms (IWTs) (which have similar connotations as their Fourier counterparts).

Muchaxo and Neves (1999) have shown that for the multiresolution representation of terrains, a relevant application in the real-time visualisation of terrains (Chapter 5), a Haar wavelet function (Starck *et al.*, 1998), is adequate. The Haar basis in one dimension simply requires averaging the values in order to obtain coarser representations, storing the details as differences between those original values (DWT). These details are used to reconstruct the function or signal to the original (more detailed) resolution (IWT). The generalisation to two dimensions (and more) can be carried out using the standard or the non-standard decomposition. In the standard decomposition, all rows are decomposed first and then all columns. In the non-standard decomposition, the rows and columns are interlaced.

Muchaxo and Neves, developed quadtrees (see Chapter 3) to store terrain data in a multiresolution fashion: a low resolution version of the data is used for the root of the quadtree and detail coefficients are organised at increasing resolutions up to the leaves. The levels of detail are stored and managed on a quadcell basis using small submatrices. The use of submatrices with different resolutions in a LOD (level-of-detail) representation is only possible if the application of wavelets to the starting matrix equals the composition of wavelets of each submatrix. In other words, the composition of children node wavelets (image and details) is equal to the wavelet of the parent and vice versa. Muchaxo and Neves showed that this is the case for the Haar basis.

A2.5 Web Sites

Links for image processing from computer vision researchers at Carnegie Mellon University are offered at (http://www.cs.cmu.edu/afs/cs/project/cil/www/vision.html).

An excellent index of links on theory and applications of wavelets is available at (http://mathsoft.com/wavelets.html).

Image processing algorithms may be found at Numerical Recipes' site (http://www.nr.com) and Netlib's Public Domain Software list (http://www.netlib.org).

Adobe Photoshop (http://www.adobe.com) is the general purpose image manipulation software which has become a *de facto* standard. An Open Source counterpart is The GNU Manipulation Program (GIMP) available free of charge (http://www.GIMP.org).

This list of sites is updated at http://gasa.dcea.fct.unl.pt/camara/index.html.

References

Gonzalez, R.C., and Woods, R.E. (1992). *Digital Image Processing.* Reading, MA: Addison-Wesley.

Jensen, J.R., (1996). *Introductory Digital Image Processing, A Remote Sensing Perspective.* Upper Saddle River, NJ: Prentice Hall.

Muchaxo, J., and Neves, J.N. (1999). *Visualisation of Large Terrains using Wavelet Functions.* Unpublished internal report, Imersiva, Monte de Caparica, Portugal.

Press, W.H., Teukolsky, S.A., Vetterling, W.T., and Flannery, B.P. (1992). *Numerical Recipes in C: The Art of Scientific Computing.* New York: Cambridge University Press.

Starck, J.L., Murtagh, F., and Bijaoui, A. (1998). *Image Processing and Data Analysis, The Multiscale Approach.* Cambridge, UK: Cambridge University Press.

Wolff, R.S., and Yaeger, L. (1993). *Visualisation of Natural Phenomena.* New York: Springer Verlag.

Appendix 3

Computer Graphics Primer

Computer graphics is the process of generating images using computers. In a simplified view, one can say that there are five relevant topics to understand that process applied to three-dimensional (3D) graphics: modelling, lighting, virtual camera, surface characteristics, and rendering.

A3.1 Modelling

Modelling objects in 3D space may follow three approaches (O'Rourke 1998):

- *Surface modelling*, where the surfaces that enclose an object define its shape;
- *Solid modelling*, where the object is defined with all the attributes of a solid (i.e., mass, density);
- *Particle systems modelling*, that deals with phenomena such as fire, clouds, mist, and pollutant plumes.

In environmental applications, surface and particle systems modelling are the most relevant (see Chapter 5). The basic geometrical shapes used in modelling are primitives. For two dimensions, primitives include a line, circle, ellipse, arc, polyline, polygon, and splines.

Surfaces can be modelled by using polygons; a process called polygon approximation (O'Rourke, 1998). Models with many polygons have what is called a high polygon count. Techniques to reduce such a count, while maintaining a reasonable degree of fidelity, are termed 'polygon thinning' or 'polygon culling'.

Splines can also be used to model surfaces (see Chapter 3 on the use of splines for terrain modelling). Spline curves are dictated by control points that determine the shape of the curves (Fig. A3.1). They are in fact mathematical formulations of curves, usually termed 'cubic splines' because they include variables raised to the power of three (O'Rourke, 1998).

There are two major types of splines: interpolating splines and approximating splines.

In interpolating splines, the spline curve passes directly through each control point (Fig. A3.1b) In approximating splines, this problem is addressed by calculating the curve

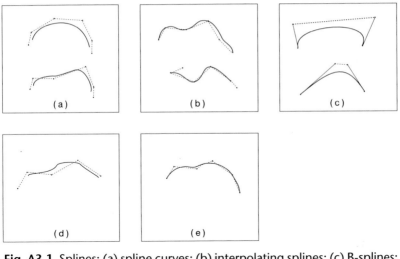

Fig. A3.1 Splines: (a) spline curves; (b) interpolating splines; (c) B-splines; (d) Bezier splines; (e) NURBS curve

Adapted from O'Rourke, 1998

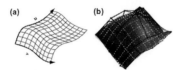

Fig. A3.2 Patch generation: (a) along u and v directions; (b) moving the control points of the patch changes its shape

Adapted from O'Rourke, 1998

that it goes near, but not directly through the control points. The most common approximating splines include:

- B–splines, where the curve does not touch any of the control points (Fig. A3.1c).
- Bezier curves, where tangent vectors are added at the end of the curve. The length and direction of tangent vectors control the shape of the curve (Fig. A3.1d).
- Non-uniform rational B-spline (NURBS) curves, where like in an interpolating spline, the curve goes through the first and last control point; however, it does not go through the intermediate control points. A set of edit points also lie on the curve, and can be manipulated to achieve more precision (Fig. A3.1e).

A curve moved in space along a straight or curved line defines a curved surface. Two splines moved along two directions (Fig. A3.2) may be used to generate a surface curve, called a 'patch'. The control points of the generating splines may then be used to change the shape of the patch (Fig. A3.2b).

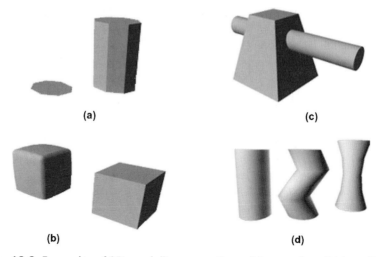

(a)   (c)

(b)   (d)

Fig. A3.3 Examples of 3D modelling operations: (a) extrusion; (b) bevelling;
(c) constructive solid geometry; (d) deformations

Three-dimensional geometric primitives include spheres, cubes, cylinders, torus, and cones. These are defined by a number of control parameters including position and key dimensions (i.e., the radius of a sphere).

There are a number of operations that enable 3D modelling from existing 2D and 3D primitives. Examples include: extrusion, bevelling, constructive solid geometry, and deformations (Fig. A3.3).

Extrusion enables the creation of 3D objects from 2D objects by pushing them straight in space. An example is the creation of 3D representations of text by pushing the letters.

Bevelling is simply the rounding of edges. With constructive solid geometry (also called Boolean operations), one can add, subtract, or intersect two objects to create a third. Finally, deformations include bending and twisting.

There are four coordinate systems considered in computer graphics: models, world, view, and display (Fig. A3.4).

Users develop models of objects in a local Cartesian coordinate system. This is called the 'model coordinate system'. Each object has its own local coordinate system.

The world coordinate system is the space in which the objects are positioned. It is also the system in which the position and orientation of cameras and lights (see below) are defined. The view coordinate system represents what is visible to the camera. It is defined by an x and y axis, with values ranging from -1 to 1, and a depth, z, specifying the distance from the camera.

The display coordinate system is based on the same principles as the view coordinate system, but the coordinate range $(-1, 1)$ of the view system is mapped on to actual x, y pixel locations. To render different scenes and display in the same window, the window can be divided into rectangular viewports (Schroeder *et al.*, 1998).

Most 3D modelling programs provide front, top, side, and perspective views. These capabilities allow viewing in 3D, actions performed in the 2D front, top, or side views.

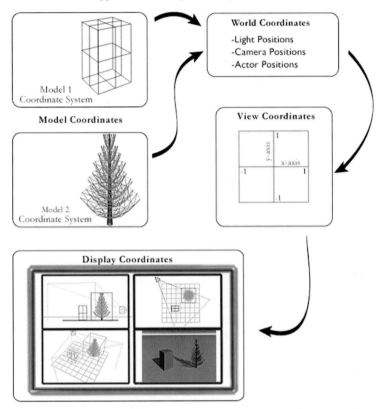

Fig. A3.4 Model, world, view, and display coordinate systems

A3.2 Virtual Camera

The 3D software applications use the metaphor of a 'virtual camera' to represent a point of view on to a 3D scene. There are two relevant parameters characterising this virtual camera:

- Its location, defined by x, y, and z coordinates in the world coordinate systems.
- The 'where you are looking' parameters (O'Rourke 1998) that may be understood as either a point in space (centre of interest) or the camera direction, when thinking in terms of the rotation of the camera.

As a real camera, one can:

- *Pan*, that is, keeping the camera fixed while moving the centre of interest.
- *Tumble*, that corresponds to keeping the centre of interest while moving the camera.
- *Track*, which means changing the camera location and the centre of interest simultaneously.

- *Dolly*, when the camera moves along the direction of projection, either closer or farther from the centre of interest.
- *Pitch*, *yaw*, and *roll*, which are rotations of the camera along the *x*, *y*, and *z* axes, respectively.

Like a real camera, one may also use different lenses to zoom in and zoom out, or obtain panoramic views. Unlike real cameras, virtual cameras may be programmed to fly through a scene.

A3.3 Lighting

Lighting models aid in viewing 3D objects on 2D displays. When you specify a lighting model, you define the location (in the world coordinate system), intensity, and colour of the light (see below). Available light models include (Rhyne, 1997):

- *Point light*, that emulates light bulbs and illuminates a scene in all directions from a single point in space.
- *Infinite light*, that mimics the effect of the sun and is composed of parallel rays.
- *Ambient light*, that produces a constant illumination on all surfaces of a 3D object.
- *Area light*, that emits a rectangular area of light.
- *Volume light*, where the illumination is restricted to a specified volume of space.

A3.4 Surface Characteristics

Surface characteristics are another element that has to be specified before rendering a 3D scene. They include the definition of the colour and shininess of a surface.

Colour and other Characteristics

The visible electromagnetic spectrum (see Chapter 2) ranges from 400 to 700 nanometres. The human eye uses three colour receptors; each of these responds to a subset of that range. Any colour one sees is the result of these three overlapping responses.

Colour coding in a computer has also been developed applying a 3D colour coordinate system. Thus, three numbers, one for each of the three primary colours, usually define the colour of an object surface: red, green, and blue (RGB) (Foley *et al.*, 1997). The RGB values range from 0 (no colour) to 255 (full colour). For instance: RGB = (0, 0, 0) represents pure black; RGB = (255, 255, 255) is pure white; RGB = (255, 0, 0) pure red; and RGB = (255, 255, 0) is yellow.

The RGB coding system is hardware based. Two other popular colour systems are cyan, magenta, and yellow (CMY) and hue, saturation, and value (HSV).

CMY is a colour system used by printers and photographers. Hard-copy devices deposit colour pigments on paper. When illuminated, each of the three colours absorbs its complementary light colour: cyan absorbs red; magenta absorbs green; and yellow absorbs blue (Murray and vanRyper, 1996). CMY colours are subtracted from white light by pigments in order to create new colours. CMY colours can be coded as a colour triple like RGB. Their individual colour values are just the opposite of RGB. Black is (255, 255, 255) and white is (0, 0, 0). As perfect black is difficult to obtain, there is a variant called CMYK. In this variant, black is treated as an independent primary colour variable.

HSV is user-oriented and can be coded in a similar fashion. The hue represents the dominant wavelength of the reflected or transmitted light from the object. Saturation reflects the difference from a given colour and a grey of equal intensity. A very saturated red will be brilliant; a less saturated red will be duller (O'Rourke, 1998). Finally, the luminance (or intensity) is the brightness of the colour and can be used to determine good contrast (Rhyne, 1997). Luminance is computed by:

$$Luminance(L) = 0.30 \times Red + 0.59 \times Green + 0.11 \times Blue$$

The luminance of black is 0, white is 1.00, and yellow, 0.89. This explains why a yellow object on a white background has poor contrast; and why the same object has excellent contrast on a black background.

Other surface characteristics result from the way the light reflects from the surface and are specified by several parameters such as:

- *Diffuseness*, a measure of how much light reflects from a surface.
- *Specularity*, meaning that a surface may also be defined by its highlights (shiny surfaces have high specularity).
- *Roughness*.
- *Transparency*.

Objects with the same colour map but with different values for these parameters may look substantially different.

Texture Mapping

The previous discussion was centred in surfaces with only one colour. Real surfaces are much more complex. Texture mapping is a process that enables one to use realistic representations on surfaces. These may be photographs, painted images, or mathematically generated effects. In texture mapping, 2D pictures are applied to the surface of the 3D model using a variety of projections which can be parameterised.

A3.5 Rendering

Rendering is the final step in the process of generating images. In simple terms, what a 3D renderer does is to look through the camera and calculate the appearance of every pixel for every frame. This implies two main processes: visible surface determination and shading.

Renderers are classified according to their method of visible surface determination. Shading algorithms can be coupled to any of these methods.

In Chapter 5, ray-tracing algorithms were mentioned. They tend to be slow and other alternatives based on hardware solutions, such as Z-buffering, are preferable if the shading effects of the interplay between objects in a scene are not a major concern (such as in most visualisation and animation efforts). Z-buffering can be outlined as follows (Wolff and Yeager, 1993):

1 Store the background colour and the maximum depth (z-value) in every pixel of the screen.
2 Transform the vertices of an object (and, thus, all the polygons of an object) into the coordinates of the screen.
3 Take each polygon in an object, and starting at one vertex, decompose the polygon into the scan lines and the pixels it overlaps. As a result, the RGB values and the z-value are obtained for each pixel the polygon includes.
4 If, and only if, the z-value at each polygon of this polygon is less than the value of depth value (z) already stored at that pixel, then store the polygon's colour and z at that pixel.
5 Go back to step 1. Repeat the procedure until all polygons of all objects have been scanned.

Z-buffering, which can be considered to fall in the general category of ray-casting algorithms (O'Rourke, 1998), has two major flaws: memory requirements and accuracy when objects are close together (Schroeder *et al.*, 1998).

Shading is the effect of light on the surface of an object and implies a colour interpolation across each polygon of an object. The colour of an object not only depends on direct lighting but also on ambient light that is being reflected or scattered from other objects in the scene. Lighting models expressed in terms of intensity values account for these effects. Shading algorithms, such as flat, Gouraud, or Phong, apply to these models.

Flat shading calculates the colour of a polygon by applying the lighting equations to the normal surface of the polygon. Gouraud (1971) shading calculates the colour of the polygon at all its vertices using the lighting models and the vertices' normals. The colour of the interior and edges of the polygon are then calculated by interpolating from one corner colour to another. Phong (Bui-Tuong, 1975) shading is the most realistic but less efficient: it calculates a normal at every location of the polygon by interpolating the vertex normals. The results of those calculations are then used in the lighting equations to determine the colour of every pixel in the polygon.

A3.6 Web Sites

ACM SIGGRAPH's site is (http://www.siggraph.org/). This site provides access to a myriad of computer graphics resources.

A helpful site is the Encyclopedia of Graphics File Formats Web page (http://www.ora.com/centers/gff/).

References

Bui-Tuong, P. (1975). Illumination for Computer Generated Pictures. *Communications of the ACM*, 18/6: 311–17.

Foley, J.D., VanDam, A., Feiner, S.K., and Hughes, J.F. (1997). *Computer Graphics: Principles and Practice.* Reading, MA: Addison Wesley.

Gouraud, H. (1971). Continuous Shading of Curved Surfaces. *IEEE Transactions on Computers*, C-20/6: 623–9.

Murray, J.D., and vanRyper, W. (1996). *Encyclopedia of Graphics File Formats.* Sebastopol, CA: O'Reilly.

O'Rourke, M. (1998). *Principles of Three-dimensional Computer Animation.* New York: W.W. Norton.

Rhyne, T.M. (1997). Internetworked 3D Computer Graphics: Beyond the Bottlenecks and Roadblocks. Unpublished notes for a tutorial presented at ACM SIGCOMM'97.

Schroeder, W., Martin, K., and Lorensen, B. (1998). *The Visualisation Toolkit.* Upper Saddle River, NJ: Prentice Hall.

Wolff, R.S., and Yeager, L. (1993). *Visualisation of Natural Phenomena.* New York: Springer.

Appendix 4

Web Addresses for Periodicals

A4.1 General Searches

Gateways to scientific and technical literature on the topics covered in this book include:

- Web of Science covers articles published in journals covered in Science, Social Sciences, and Art and Humanities Citation Indexes (http://www.webofscience.com). It displays abstracts for most articles published in journals reviewed in those indexes.
- Engineering Information (http://hood2.ei.org) provides access to engineering indexes and has an active service on relevant engineering news, among other services.
- Online Computer Library Center (http://www.oclc.org) provides access to a variety of services including abstracts and papers of journals and conference proceedings stored in the British Library.
- Ovid (http://www.ovid.com) is a gateway to abstracts and complete papers of services such as *ABI/Inform* (economics, management, information technologies), *Cambridge Environmental Science and Pollution*, *INSPEC* (electronics, physics, information technologies), and *Dissertation Abstracts*.
- SilverPlatter (http://webspirs.silverplatter.com) is a gateway to databases such as *Geobase* (geography and ecology literature), *GeoRef* (geosciences literature), and *Water Resource Abstracts*.
- Uncover Web (http://uncweb.carl.org) provides a document delivery system to articles published in over 17,000 journals.

The Web sites of relevant magazines and periodicals are listed below. Most provide free access to abstracts. Many provide full text with paid subscriptions.

A4.2 Magazines

Science and Technology

Scientific American (http://www.sciam.com)
Technology Review (http://www.techreview.com)

Environment

National Geographic (http://www.nationalgeographic.com)
The Ecologist (http://www.theecologist.org)

Spatial Information Systems

Earth Observation Magazine (http://www.techexpo.com/toc/eom.html)
GeoEurope, GeoWorld, and Mapping Awareness (http://www.geoworld.com)

Computing

Byte (http://www.byte.com)
Computer Graphics World (http://www.cgw.com)
Dr. Dobbs (http://www.ddj.com)
PC Magazine (http://www.pcmagazine.com)
Web Techniques (http://www.webtechniques.com)

Information Society

Red Herring (http://www.redherring.com)
Wired (http://www.wired.com)
Technological news on line (http://www.cnet.com)

A4.3 Journals

Science

Nature (http://www.nature.com)
Science (http://www.aaas.org)

Environment

Advances in Water Resources (http://www.elsevier.com)
Atmospheric Environment (http://www.elsevier.com)
Ecological Modelling (http://www.elsevier.com)
Ecology (and other journals of the Ecology Society of America) (http://esa.sdsc.edu)
Ecosystems (http://link.springer.de)
Environment Management (http://link.springer.de)
Environmental Conservation (http://www.journals.cup.org)
Environmental Impact Assessment Review (http://www.elsevier.nl)
Environmental Modelling & Software (http://www.elsevier.nl)
Environmental Monitoring and Assessment (http://www.wkap.nl)

Environmental Science and Technology (http://pubs.acs.org/)
Journal of Air and Waste Management Association (http://www.awma.org)
Journal of Computing in Civil Engineering, ASCE (http://www.asce.org)
Journal of Ecology (http://www.jstor.org)
Journal of Environmental Engineering, ASCE (http://www.asce.org)
Journal of Environmental Management (http://www.idealibrary.com)
Journal of Environmental Psychology (http://www.idealibrary.com)
Journal of Environmental Systems (http://baywood.com)
Journal of Hydraulic Engineering, ASCE (http://www.asce.org)
Journal of Hydrologic Engineering, ASCE (http://www.asce.org)
Journal of Industrial Ecology (http://mitpress.mit.edu)
Journal of Infrastructure Systems, ASCE (http://www.asce.org)
Journal of the American Water Works Association (http://www.awwa.org)
Journal of the Water Environment Federation (http://www.wef.org)
Journal of Water Resources Planning and Management, ASCE (http://www.asce.org)
Landscape and Urban Planning (http://www.elsevier.nl)
Landscape Ecology (http://www.wkap.nl)
Sensor Review (http://www.mcb.co.uk/sr.htm)
Urban Ecosystems (http://www.wkap.nl)
Waste Management and Research (http://www.iswa.org)
Water Research (http://www.elsevier.nl)
Water Resources Research (http://www.agu.org)
Water Science and Technology (http://www.elsevier.nl)

Spatial Information

Cartography and Geographic Information Systems (http://www.survmap.com)
Computers & Geosciences (http://www.elsevier.nl)
Computers, Environment and Urban Systems (http://www.elsevier.nl)
Environment and Planning B (http://www.pion.co.uk/ep/)
Geographical Analysis (http://thoth.sbs.ohio-state.edu/faculty/okelly/ga.html)
Geoinformatica (http://www.wkap.nl)
IEEE Transactions on Geoscience and Remote Sensing (http://www.ieee.org)
International Journal of Geographical Information Science (http://www.tandf.co.uk)
International Journal of Remote Sensing (http://www.tandf.co.uk)
Photogrammetric Engineering and Remote Sensing (http://www.asprs.org)
Photogrammetry and Remote Sensing (http://www.isprs.org)
Remote Sensing and the Environment (http://www.elsevier.nl)
URISA Journal (http://www.urisa.org)

Modelling

ACM Transactions on Modeling and Computer Simulation (http://www.acm.org)
Decision Support Systems (http://www.elsevier.nl)
European Journal of Operational Research (http://www.elsevier.nl)
Evolutionary Computation (http://mitpress.mit.edu)
IEEE Systems, Man and Cybernetics: Applications and Systems (http://www.ieee.org)

IEEE Transactions of Evolutionary Computation (http://www.ieee.org)
Interfaces (http://www.informs.org)
Journal of Theoretical Biology (http://www.academicpress.com/jtb)
Management Science (http://www.informs.org)
Operations Research (http://www.informs.org)
ORSA Journal of Computing (http://www.informs.org)
Simulation (http://www.scs.org)
System Dynamics Review (http://www.interscience.wiley.com)
Transactions of the Society for Computer Simulation (http://www.scs.org)

General Computing

ACM Computing Surveys (http://www.acm.org)
Communications of the ACM (http://www.acm.org)
IEEE Computer (http://computer.org)
The Computer Journal (http://www.oup.co.uk/computer_journal)

Graphics and Image Processing

ACM Transactions on Graphics (http://www.acm.org)
Computer Graphic Forum (http://www.eg.org)
Computer Graphics (SIGGRAPH Proceedings) (http://www.siggraph.org)
Computer Vision, Graphics and Image Processing: Graphical Models and Image Processing
   (http://www.apnet.com)
Computers & Graphics (http://www.elsevier.nl)
IEEE Computer Graphics and Applications (http://computer.org)
IEEE Transactions on Image Processing (http://www.ieee.org)
IEEE Transactions on Visualization and Computer Graphics (http://computer.org)
Journal of Electronic Imaging (http://www.spie.org)
Journal of Graphics Tools (http://www.akpeters.com)
Journal of Visual Communication and Image Representation (http://www.idealibrary.com)
Journal of Visualization and Computer Animation (http://www.interscience.wiley.com)
The Visual Computer (http://link.springer.de)

Internet

IEEE Internet Computing (http://computer.org)

Databases

ACM Transactions on Database Systems (http://www.acm.org)

Multimedia

IEEE Multimedia (http://computer.org/multimedia)
IEEE Transactions on Multimedia (http://www.ieee.org)

Multimedia Systems (http://link.springer.de)
Multimedia Tools and Applications (http://www.wkap.nl)

Virtual Reality and Artificial Life

Artificial Life (http://mitpress.mit.edu)
Presence (http://mitpress.mit.edu)
Virtual Reality (http://link.springer.de)

Interface Design

ACM Transactions on Human-Computer Interaction (http://www.acm.org)
Computer Supported Cooperative Work (http://www.wkap.nl)
Human-Computer Interaction (http://www.erlbaum.com)
Interactions (http://www.acm.org)
International Journal of Human-Computer Systems (http://www.idealibrary.com)
SIGCHI Bulletin (http://www.acm.org/sigchi)

Subject Index

Author Index